Highland Homespun

Highland Homespun

MARGARET LEIGH

ORIGIN

This edition published in 2018 by
Birlinn Origin, an imprint of
Birlinn Limited
West Newington House
10 Newington Road
Edinburgh
EH9 1QS

www.birlinn.co.uk

First published by Birlinn Limited in 2012

ISBN: 978 1 91247 609 1
eBook ISBN: 978 0 85790 298 6

British Library Cataloguing-in-Publication Data
A catalogue record for this book is available
from the British Library

Typeset by Hewertext UK Ltd, Edinburgh
Printed and bound by Clays Ltd, St Ives plc

TO
G. AND A.M. FORBES OF FERNAIG

Contents

1

Foreword: Hail and Farewell

CURSED as we are with universal education, there are few people who have not at one time or another dabbled in literature. I am not speaking of professional writers, but of the numerous amateurs who have taken to writing either to fill the leisure of retirement or as a distraction from a business that is too dull, unprofitable, or depressing to claim their whole attention. Even the inarticulate race of farmers has been driven into print by the slump in agriculture, often to find that books pay better than the farm; for the British public, at last repenting of its long neglect of rural problems, seems to have developed an almost passionate interest in the land.

This book is about farming, but not perhaps about the kind of farming that most of us have known, and most certainly not the kind that must (alas for the necessity) prevail in the future. The farm I have called Achnabo is worked very much as it would have been in the days of Abraham. All operations are performed on a small scale and by hand, not for the sake of theory or as an experiment in apostolic simplicity, but because in our special conditions of land, labour, and climate, even a moderately mechanised system of farming would not pay. I have no axe to grind in the matter, and in praise of West Highland methods shall only say that the life, though laborious, is pleasant; and for young people intending to farm, no matter where or how, it provides an unrivalled training in the primitive manual operations on which their mechanical substitutes are based. It is of course possible to work a power loom intelligently without any knowledge of hand-loom weaving – it is done every

1

day by thousands of mill-hands; but for a thorough understanding of the whole process of weaving, the man who can operate a hand-loom is in a greatly superior position.

There has been much discussion of late on the question of preparing boys for the Navy and Merchant Service. Some are in favour of a preliminary training in sail, on the ground that in no other way can resourcefulness and intelligence at sea be so thoroughly developed. Others argue that for boys who will serve on ships propelled by machinery, a training in sail would be an expensive waste of time. It is, however, significant that the Scandinavian countries, renowned in history for their skill in seamanship, have shown most interest in the old tradition, so much so that in Finland no one can qualify as an officer in steam who has not served an apprenticeship in sail. In its constant demand for resourcefulness, initiative, and a skilled weather sense, the primitive unmechanised farm is like a sailing-ship, and no one, I believe, can make proper use of agricultural machinery who has not first mastered the routine of horse and hand cultivation, and studied the habits of plants and animals with a detailed accuracy that is possible only with personal attention and small numbers.

The small hand-worked farm has another advantage – it teaches the beginner how to make do with what he has already, or if he must get something new, to contrive it himself out of any odd material lying to hand. A common fault in agricultural education is to provide the student with completely up-to-date, scientific, and fool-proof equipment of a kind that would never be found in practice except on the model farm of a millionaire. The natural result is that the beginner, fresh from his vision of harvest combines and white-tiled dairies, spends much of his working capital on unnecessary improvement of buildings and implements. Every agricultural college should have an improvisation course, with an instructor from the back-blocks of Australia or Canada, who would teach students how to use up the quantities of old iron, wood, bits of wire-netting, and so forth, that lie upon the scrap-heap of every farm. If this were done, there would be fewer failures among agricultural students who take up farming on their own account.

I have used the word 'method' once or twice, but the beauty of Highland farming is that there is little or no method about it. This is not quite as feckless as it sounds, for the weather is so variable and the people so indolent that the most sensible and profitable thing is to throw method overboard and swim with the stream. The best-laid plans may be upset by a sudden change in the weather or the non-arrival of an indispensable casual worker, and it seems a waste of time to make them. So this book, like the course of our work on the farm, drifts idly along, following, it is true, the procession of the seasons, but otherwise moving here and there in pursuit of interest, just as a young dog on a walk follows his master's general direction, but continually breaks away to investigate some new and seductive scent.

It is a book written at the kitchen table of a Highland farm, and describing the life that is lived there. But it is not a book about the Highlands. Of these there is an ever-increasing number, some good, some bad, but nearly all well illustrated, and perhaps justified by the beauty and interest of the pictures alone.

The glens and lochs of the north-west have been opened up to a large and intelligent public in the south, which demands topographical books of all kinds, from the chatty guidebook to the technical work on natural history, philology, or folklore. Thousands of people, though they may never have seen them, can distinguish the various peaks of the Cuillins of Skye, can name the varieties of seabirds found on the crags of St Kilda, or of wild flowers blooming on a machair in Uist. They know the meaning of Gaelic place-names, can sing Hebridean songs, are familiar with the ancient legends. A certain amount of this interest in Celtic things is literary or artistic, a transient fashion, sometimes merely a pose. But I am sure that it goes deeper than this.

Some years ago I was told of a London typist who, fired by the works of Fiona Macleod, spent her savings on a visit to Iona, where she was overtaken by the tide and drowned. A silly schoolgirl escapade; and the writer who inspired it is, to my plain Saxon taste, intolerably sentimental. But behind this silliness is the desire, strong and wholly reasonable, to escape from the stultifying bondage of our commercial civilisation. And it is this desire that makes people delight in reading of the wild country that still lies unspoilt, for our pleasure and refreshment, in the far north-west. Changing conditions of life have changed our values. A thousand years ago, the invaders of Britain preferred the flat and fertile south and east, and their choice was a sound one, for a rich living was to be found there. The vanquished were driven into the hills of the Atlantic seaboard, where there was little land worth cultivating. As late as the eighteenth century, Englishmen whose work took them to the Highlands pitied themselves for being condemned to a miserable exile. If they wrote books about the country, they were catalogues of complaints. Now we have changed all that: the horrors of the eighteenth century have become the beauties of the twentieth. No glen is too gloomy for us, no crag too beetling, no loch too grim, no strait too stormy. But if I may digress for a moment, are we sure, when we smile at the miseries of Johnson or Burt, that we are in a position to judge? For us, a visit to Skye means a comfortable journey by corridor train and steamer, or by luxurious car upon tolerably well-made roads, with a good bed at the end of it. No wonder that gloomy Sligachan has lost its terrors, and is no more than a fine piece of scenery to be glanced at with pleasure before ordering our dinner at the hotel. For them, it meant difficult and dangerous travelling on horseback through bogs and thickets, among rocks and precipices, or in small boats, at the mercy of wind and tide, their only shelter whatever hovel they happened to find when darkness overtook them. If at a blow we were to lose every amenity of modern travel, and be condemned to the use of ponies and rowing-boats, our attitude towards the sublime in scenery would, I feel sure, be rather different.

To return to our argument. Here, it seems, is the only place where you can walk all day without seeing anyone, wear what you like, work when and how you like (or not at all if you don't like), never make plans, never carry a purse, never lock your door, leave things lying about because people are too honest or too indolent to take them. Blessed land of freedom and ease! No wonder that many of us, provided it is not for good, want to live in it, and that most of us want to read about it. For in reading we get much of the delight of travel and adventure without the danger and discomfort. I knew an old lady who dared not face the mildest cow; yet her favourite reading was of man-eating tigers. Is this one of life's little ironies, or an example of the law of compensation? I neither know nor greatly care. Only I am sure that it is pleasant to sit at home and read of things remote and dangerous. Pleasant enough, until there is a cry of 'The cows are in the corn!' or 'The potatoes have boiled dry!' And then stark tyrannous reality swoops down upon us, and the exploits of gallant explorers or blubbery whalemen fade into the limbo of things forgotten and infinitely unimportant.

Why not, then, another book about the Highlands? Because, though I have lived for ten years in this seductive land, I know very little about it as a whole. These ten years have been spent in one district, doing my own work and living my own life. Most of the notable things, the things that people write and dream and talk about, are unknown to me. I have never seen the surf breaking on St Kilda, nor the flood-tide covering the fords of Benbecula, nor Loch Maree, nor Glencoe, nor the sun setting over Morar Bay. Of Gaelic I have but a few words, and only the haziest idea of the stories of Deirdre and Cuchullain. And there is worse. A topographical book should be written either by a native, to whose inside knowledge and natural sympathy many rare and interesting things are revealed, or by an observant stranger, who, provided he has sufficient insight, may often enjoy a clearer and less biased point of view. But the mere inhabitant, who is neither native nor visitor, has lost the latter's freshness of outlook without gaining the intimacy of the former.

And so what is offered here is not a book about the Highlands, but an unsystematic bundle of reflections upon life on a Highland farm, where the workers, scenery, and temperament of the place give to the life lived there its own peculiar flavour. If I dare write of these things, it is because I have herded cows on Highland hillsides, hewn timber in Highland woods, worked side by side with crofters and shepherds, not as a social or literary experiment, nor with any Tolstoyan theories about the excellence of peasant life in itself, but just as it happened to fit best in the doing of my own work. This is a point I specially want to stress, because this book may seem to be the work of an intellectual temporarily surfeited by civilisation, who takes a spell of country life for medicinal purposes, as one might take a dose of salts after too full a meal. The idea is sound enough, but the books written by these people have something a little transitory and unreal about them. Today we are in Thule; but tomorrow we shall be (thank heavens!) back in Bloomsbury or Montmartre, writing in quite another strain. The cure is effected and we have no further need of cabbages or cows to soothe our frayed nerves or stimulate our jaded fancy. If forced to theorise, I would certainly admit that the simple life is the only completely rational life, because it is the only completely natural one; and man is superior to the beasts not in setting Nature aside, but in learning to work harmoniously within the limits it lays down; and this by no mere animal instinct, but by conscious and intelligent acquiescence. But beyond admitting this main principle, I have no taste for further argument. The simple life is desirable only as long as it is unself-conscious. The moment we begin to theorise about ourselves, we cease to live simply. Work is good, but theories of work are bad, if for no other reason than because the theoriser has to stop working in order to be able to talk.

Almighty God Himself cannot put back the clock. Neither can we, by force, preaching, or example, call back the past, and reduce our huge population to a level which would enable us to become a self-supporting agricultural community. We have reached an unpleasant stage of evolution. We may, as optimists suppose, be heading straight for the millennium. If so, it must be a madman's paradise in which ever-increasing motion will have produced, as in some cosmic traffic block, a state of perpetual stagnation. But in any case we must go forward, not back. We cannot, if we would, abolish motor cars, blast furnaces, power looms, advertising, wireless, cinemas, jazz bands, and all the whirling clanking pandemonium that is called civilisation. It is doubtful if any of these have added one iota to the sum of human happiness; indeed there is even reason for thinking the reverse. But they have come to stay, at all events for a time; and if people want them they must have them. We can only believe that the sanity and common sense of ordinary people will in the end prevail, so that, unless too far hypnotised by advertising stunts and the daily press, not to mention a dozen other agencies of corruption, they will demand a quieter and more satisfying way of life than anything our speedomaniacs have to offer. Happiness is certainly, as we were told long ago, the accompaniment of activity, but not of an activity so frenzied that it leaves no time for enjoyment and reflection. In Heaven's name, allow us to chew our cud.

In the meantime, while we await an opportunity to build Jerusalem in England's green and pleasant land (so much less green and pleasant than in Blake's time, for he only foresaw as a prophet what we endure as a sordid reality, and all through our own fault), a few of us, far from cities with their grimy work and tinsel amusements, may try the experiment of living more simply and more slowly. But let us lie low, and be humble about it. There is no use setting up as a chosen people, especially beloved of God and assured of some particular salvation. Not merely shall we be disliked, but we shall miss the happiness that only comes by stealth to the unaware. And in any case it will give us a salutary shock to reflect that our freedom is possibly bought with the profits earned by producers of petrol, gramophones, and motor tyres. We are hypocrites. Perhaps: but we need not be. We can sell our own Shells and Dunlops, and invest our money in livestock. Prices are low just now, and with the possibility of a rise in the near future, the gamble might be worth making. What about it?

Before plunging the reader into the midst of our unsystematic life and work, it is only fair to introduce him to the farm itself and the way in which I came to occupy it.

Achnabo stands in an open position about 350 feet above the sea, with a hundred-odd acres of enclosed land and 600 more of woods and rough grazing. There had always been a few crofters on the spot, but about seventy years ago the proprietor cleared, fenced, and drained more land, and made the whole place into a home farm to supply his castle with produce. The enclosed fields were then in fine condition and carefully tilled in rotation. Machinery was not used, and armies of local people were employed to cultivate and harvest the crops. Not a rush, not a weed, was to be seen anywhere. Old people can still recall this glory, but now it has become an almost incredible legend. The fortune that wrested our land from the wilderness has been dissipated, the estate is in other hands, and Nature is rapidly recovering her lost dominion. The sleek fields bristle with rushes, thistles, and tansy. The flats have become sour and water-logged through want of lime and neglected drains. Fencing posts have crumbled to dust, wires are slack, twisted and broken, gates lie flat or hang ready to fall. The houses and buildings have been recently renovated, but everything else is hurrying back to the wild as fast as it knows how. And when we are gone it will probably go a bit faster, for unless the land is divided into small holdings, I doubt if it will be farmed independently again. For there is no hill ground for sheep, and no accessible market for the products of intensive farming; so that, unless the price of store cattle should rise to heights unknown today, the farm, with all its burden of long neglect, cannot be made to pay its way. It will probably be annexed by some powerful neighbour, to support a few sheep and a dozen or so of out-wintering cattle.

My tenancy of Achnabo began at the May term, 1933. The farm had stood for some time un-let, and I got it on very generous terms. My first idea had been to use Achnabo as a preliminary home for Poor Law children who would ultimately be passed on to the Fairbridge Farm School in Western Australia. This scheme, though encouraged by the Child Emigration Society itself, had to be abandoned, because the authorities would not send children to a district beyond the ordinary reach of their inspectors. I was then no stranger to the Highlands; we had come to settle there in 1925, and returned to England only for short occasional holidays. I had become more and more absorbed in agriculture, and the four years previous to my entry into Achnabo were spent on the farm of my friend at Strathascaig, whom we always call the Laird, because on his territory of 6,000 acres he does exactly as he likes, and has a paternally autocratic way of dealing with everybody and everything. He and his wife have been and always will be the dearest of friends and neighbours, and without their constant help and advice we should often have been in a sorry mess. Other neighbours are few: the nearest crofting village is nearly a mile away, and the nearest house is out of sight.

At the moment, the stock consists of seven cross-bred milk cows and suckers in the byre, six 'Belties', or out-wintering Belted Galloway in-calf heifers, nine calves of six months and over, a pedigree Aberdeen Angus bull, a horse, about a hundred head of Light Sussex and cross-bred poultry, and two dogs. This year we had nine and a half acres of oats, a quarter of an acre of potatoes, about twelve acres shut up for hay, and the rest pasture. The regular permanent staff is small enough – myself, and Peter, a public school boy of eighteen. In the summer, when there was no time for housework, I had Flora, a gnome-like little being from the village below, with a boundless devotion to everything and everybody on the farm, and an equally boundless indifference to time and order. The poultry were looked after by Mr Gordon, the tenant of the farm cottage, and during the haymaking and harvest season we employed another public school boy in the house who worked for his keep.

Apart from questions of profit and loss, Achnabo must be one of the most beautiful places in Scotland, and neither picture nor description can give an adequate idea of its charm. The photograph at the beginning of this book shows a view from a neighbouring hill, at the foot of which the farm with its fields and buildings lie spread like a map. In the midst is the L-shaped stead-ing, the upright of the L consisting of a line of buildings – the cart-shed with granary above, the cow-byre and bull-house, with the Gordons' cottage at the far end. At right angles to these, back-ing on the stackyard, is the sawmill, once built to house a circular saw for estate work, but now used as a good substitute for a Dutch barn. At a lower level can be seen the heifer shed, stable, and long barn, and away to the left, secluded in its walled garden, the white farmhouse, an ugly little box with an asbestos-tiled roof, but easy to work and for the time being our own. Round about lie the hayfields and cornfields, the lower pastures, and the rushy field where we keep our poultry, ringed by magnificent drystone dykes and the wire fences on which we dry our hay: and beyond the fields, the woods of oak and pine, and beyond them again the sea, and Raasay and Crowlin and the hills of Skye. A view to dream about, defying words or camera or pencil, but by good fortune the background of our daily life – of washing billowing in the gale, and calves fed by guesswork in the dark; of hay first cut in hope with new scythe-blades flashing in the sunshine of July, and moistly gathered by weary arms in the storms of late October. To the north-east, and out of the picture, is a group of pasture fields, well sheltered by woods and sloping to the shores of a large freshwater loch, useful for grazing, but with value much dimin-ished by the ruinous state of the fences.

Every time I look upon this gracious scene, at whatever season of the year, I see some fresh beauty to delight the eye, and bind the heart more closely to a place that must inevitably be aban-doned, either to solitude or to someone who need not make his living from the land. Left to ourselves, we would both stay here for good; but driven by necessity, we must seek some other farm where the work will be duller and more organised but the profits greater. Here there is plenty of land, but half of it useless because the fences are in ruins, so that you cannot keep your own beasts in or other people's out. The Highland farmer depends for his living mainly on sheep; but we cannot keep sheep because there is no hill ground for them. With good fences, an adequate stock of cattle and poultry, security of tenure, and all the work done by the farmer and his family, the place could be made to pay. But meantime the fences are not there, nor the security of tenure, nor the farmer's family. However hard-working she may be, the single woman is crippled by a comparatively large labour bill.

I suppose our next farm will be in the south. We are both English, and unless something very favourable turns up, shall in due course return to our native land. Exclusive Highland ladies, some of them half Sassenach, have questioned the right of an Englishwoman to settle in the Highlands. Why do so many Scottish people settle in Sussex? Presumably because they like it, and I have never heard any Englishman raise an objection. England is of course notoriously receptive of queer importations from abroad: but why should not Scotland be equally hospitable? If I like to be rained and blown upon all the year round on a bleak Highland hillside, instead of basking at Brighton with *émigré* Macgregors and Mackintoshes, why shouldn't I? I came to the Highlands because I love the combination of mountain and sea, and that changing and delicate beauty of atmosphere which is possible only in a rainy climate; because I love space and solitude, and hate the noise and vulgarity which motor traffic and modern building development have brought to the land of my birth. I have lived for ten years in the Western Highlands, and would like, before I go, to give a picture of life as we have lived it on the farm, as remote from the sporting and social amenities of the rich and fashionable as from the childish enthusiasms of nationalist pipers and harpists. Unfortunately I have no Gaelic, but through working beside them and sharing their farming interests and anxieties, I have come to know something of the crofters and shepherds, their lives and problems. It has all been fresh, strenuous, and amusing, and, looking back upon our tenancy, I would not have missed a minute of it.

2

February: The Farmer's Year Begins

THERE are people in the south who seem to think that in midwinter we have perpetual darkness, as they do at the Pole. Actually there is more daylight than in London, because of the rare purity of the air; and the situation of Achnabo is so open that we do not even lose the sun. But it is true that in the darkest months there is not much done on the farm. Autumn and winter ploughing is of little use, as the torrential rains of December and January would flatten the furrows, leaving a hard pan that must be broken again in the spring. Also, there is hardly any frost to break up the soil. Thus the farmer's year does not really begin till February, when the days are longer and the weather as a rule a great deal drier, and sometimes too a great deal milder. Often the first month of the year comes in with gales and floods, and goes out in a dream of stillness, so that in early February, in these northern latitudes, the whole world is at peace, not in death, but with the certain hope of life's renewal. A day of this kind, now several years past, remains in my mind as if it were yesterday. There was no actual sunshine, but the soft clouds were shot with opalescent gleams of palest blue, rose, and gold. The distances were veiled in haze, so that the hills of Skye were barely visible; in front of them a misty beam of light, such as in childhood I thought to be the eye of God, slanted down upon the sea. A faint breeze from the south-east brought with it a perfume which is peculiar to this wind but common to all seasons, the scent of the wilderness, the stored fragrance of a million acres of hill.

We were helping the Laird to prune the apple trees at Strathascaig.

The orchard, fenced with hedges of beech and wild plum, sloped to a low stone dyke beside the river. The season had been unusually free from frost, and in this sheltered spot the sward, consisting mostly of tufted hardy cocksfoot, was still green. Here and there were clumps of snowdrops in bloom and daffodils in bud. The wild plum bore a few chaste blossoms, gleaming like pale stars upon the dark and spear-like shoots. Blackbirds and thrushes sang in the garden; the shrill notes of the robin were heard from the byre. The geese, who prefer to mate in the water, had gone down to the ford with much harsh clangour and flapping of wings. At the hives a few bees were making preparations for spring cleaning. Below the orchard the river purled sleepily over the stones. On such a day we could gladly pass from walking to standing, from standing to sitting, from sitting to lying, and then by slow painless transitions from dreamy drowse to deepest sleep, and thence to death, as a boat drifts with the tide, spilling our souls upon the milky air, with bodies left to make that green grass greener.

Peace absolute and, as it seemed, eternal. And yet we knew that many a lengthening day of piercing cold and blizzard, with blasted plants and dying lambs, lay between us and the solemn twilights of June. There was something uncanny about this soundless truce of Nature.

Yes, and in any case the fine weather would not last long, and we had better make the most of it. The orchard was a charmed enclosure, a place of seclusion in which to smell flowers and think poetically and at last fall asleep. But outside the air was full of the pungent odour of dung. To many people this scent is unpleasant. Now neither perversity nor a love of all things connected with the land will make me try to prove it a delightful perfume: it is not. But it is stimulating and suggestive, it evokes a host of memories of vivid manly things, like the smell of the seashore at low tide, of paint and tar on boats, of horse and harness in the hot sun. It is of the essence of spring; the stirring of life in the soil, the germination of seeds, the first springing of the tender young corn are all called up by a single whiff of this most pregnant aroma.

13

The drowsy scent of hawthorn is full of spring's languors, its careless and fruitful embraces; but the smell of dung means labour and stubborn conquest, and the slow and thrifty cycle of the farmer's year, by which one natural process is made to serve another in the interests of man. Thus beasts restore to the soil some part of the goodness they have taken out, a process which culminates in the arable farmer's fattening of cattle to manure his crops. Shovelling dung, loading it on carts, and flinging it about on the fields may seem a dirty and unworthy occupation, but it is just as important as the ploughing which the townsman considers the dignified symbol and crown of all husbandry. The Romans who summoned Cincinnatus to be consul found him at the plough. Had they come a month earlier, they might have seen him spreading dung.

The arrival of a mild and sunny spell in early February was the signal for an attack on the long dung-heap which faced the byre and (alas for our clean-milk enthusiasts) the dairy. After breakfast Peter went down to the stable to harness Dick. This beast was one of our daily crosses. A fine upstanding black horse, of about 15½ hands, something between a Clydesdale and the heavier type of Highland pony, he was a powerful creature, useful for all the labours of the farm, but spoilt by an unwilling spirit and a most inordinate greed – the walking belly, as Peter used to call him. He was expensive, too, for when I bought him the grass sickness was very prevalent, and horses scarce and dear. Also he cost a good deal to keep in winter, and lost an incredible number of shoes. Until in despair I took to having ready-made shoes put on by one of the crofters, someone had to take him to the smithy every two or three weeks. The nearest smithy was nearly four miles away, and the most reliable nearly twelve, while the maximum pace to be got out of our unshod Dick was about two miles an hour. The nearest smith was often drunk, sometimes completely dead to the world and sometimes merely silly, and even at his soberest made a poor job of it and charged eleven or twelve shillings a set. There was always a risk of finding him either too tipsy to do his work or away celebrating somewhere else, and one might have to carry on for another six miles. Here the smithy itself was near the road, but the smith's house was on the farther side of a narrow sea loch, so that you had to shout and whistle until the smith happened to hear you, when he rowed across and did the job, and very well too. It is a long time since I took a horse to that smithy; on that occasion the old man was ill, and the work was done by his son, who had come home on holiday from Edinburgh, where he was a medical student. Later on I heard that this lad, who was an excellent smith, had abandoned medicine for the anvil. And a very good thing. Scotland is overrun with doctors and nurses, and unless the health of the population greatly deteriorates, most of them will find themselves unemployed, while people able and willing to do the ordinary work of the world are unobtainable.

To go back to the dung. As Peter opened the door of the stable, Dick turned to stare at him with an insolence all his own; and then when the collar was taken down and he realised that work was on the programme, he began to bite the manger in his fury, and then made an onslaught on the collar, until, when that was pushed securely home, he owned himself beaten, and merely kicked once or twice while Peter was wrestling with the stiff rusty buckle of the decayed belly-band.

I should have been ashamed of my harness. It was never cleaned nor polished; the straw lining of the collar was protruding through holes made by some enterprising rat; the bridle was mended with string, and the reins, having in the shortage of rope been made to serve for this emergency or that, were knotted in half a dozen odd places. The saddle was a little better, as the padding had once been neatly repaired by the shoemaker with a piece of leather adapted from one of his old aprons. When I first came to the farm, I vowed that my harness was going to be properly kept. But in time laziness and the custom of the country prevailed. Also the harness was a job lot, bought for a lump sum along with a crofter's cart and a set of plough chains, and I concluded that it would not repay the labour. I hope that with a bit more string and careful treatment it may see us out.

The horse was harnessed and we got to work. The dung, soaked by continuous downpours and nearly unmixed with litter, was sloppy like porridge and very heavy to handle. Dick was in a bad mood. To make him stand, we had to tie up his head in an intricate fashion, and even so he would try to bite anyone passing in front of him. We filled the cart and rumbled off to the sawmill field, leaving deep ruts in the soft ground. Peter shovelled the dung off the cart while I spread it. The sun was shining brightly. There was no wind, and we soon got hot enough to work up a thirst, which was stayed with oranges. About half past twelve, when we had taken out several loads and were thinking of unyoking for dinner, a brilliant figure was seen approaching. It was Angus, Flora's brother, resplendent in a multi-coloured pullover of which the prevailing colour was a stark Reckitt's blue. He had lately been paid off from the forestry, and found time hanging heavily at home, where his father, uncle, and mother were more than able to cope with the work of the croft. Moreover Angus, being turned twenty and rather fancying himself with sheep and cattle, resented being ordered about by his elders. So he had come up to see if there was anything doing with us; he did not want any pay, but would give Peter a hand if we needed him. He had already had dinner, so we joyfully handed over the cart and went in.

In spite of conversational interludes with people on the road, carried on at long range from the cart in resonant Gaelic, he managed to take out several loads before Peter joined him. But he had no eye for judging distances, and the heaps of dung were of various sizes and spacing, so that when we came to spread them, there were many patches of grass that remained undunged, and had to be filled up later. I was always glad when Peter worked with one of the natives; they liked him, and told him all kinds of interesting things they would not have breathed to me, being doubly secluded as woman and boss. Some of these Peter retailed to me, but I daresay he kept a good many to himself.

The sight of Angus' pullover made me reflect that at no very distant date the native handspun tweeds and woollens, so durable in wear and pleasing to the eye, so perfectly adapted to the climate and the lives of those who originally used them, will be worn only by the faddist or the rich Sassenach on holiday, while crofters and shepherds dress in ready-made slops from Manchester. Catalogues from purveyors of cheap factory clothes can be seen by the hundred in Skye and the Outer Islands, as these firms have an enormous postal business, so that grey flannel trousers and jazz overalls are rapidly becoming the daily wear of the farthest Hebrides. Nothing can stem this tide; it is part of the flood of dreary, graceless uniformity that is sweeping over the world, reducing everyone and everything to the same undistinguished level of vulgarity. Against this force the efforts of national enthusiasts and lovers of beauty and the old traditions will be of little avail. Stimulated from outside, tweeds will still be spun and woven by hand, but not to be worn by the people themselves. What is the influence of An Comunn Gaidhealach compared with that of the big commercial organisations, playing on the natural desire of the young to be in the swim?

Angus' mother was a fine spinner, and he had more than one suit of the tweed she had spun. But Flora never wore such things, not even hand-knitted stockings. She had a wonderful array of flimsy and gaudily patterned dresses and aprons, each one uglier than the last and none of them mercifully in the least likely to last for more than a few weeks. As might be expected, the girls are the greatest innovators; many go to service in the big towns, and return in fashions that excite the wonder and envy of their sisters at home. Some years ago, when on holiday in the Outer Islands, I went down Loch Roag in Lewis in the mail motor launch. The passengers were mostly tinkers going to Bernera, but there was a bevy of elegantly dressed girls with smart composition suitcases, obviously crofters' daughters in service returning home on holiday. These damsels were precariously landed at various points on jutting spits of rock, slippery with seaweed and encrusted with barnacles, to await the arrival of someone to help them home, for their heels were too high to allow them to walk far across the wet and stony wastes of Lewis. Luckily for them, the loch was calm and the weather fine.

To go back to the dung for a moment. The application to the sawmill field, which was to be laid up for hay, was very effective, and we had a much heavier crop in consequence. The rest of the dung was put later on to a field from which a crop of oats had been taken the year before, and the grass sown with it left down. On this rough and lightly cultivated land, first year grass is often too thin for mowing, and is better grazed. But the dung was rather a mistake, as the herbage grew too rankly to be palatable, so that the field was eaten down in patches. The horse manure from the stable was carted on to the other hay field, where it produced an early growth of grass, but owing to the excessive rainfall of July and August, this part of the field was badly laid and very difficult to cut.

About this time it became obvious that unless we bought another cow we should be short of milk. Besides ourselves we were supplying the Gordons and the Rattrays, and I liked to have plenty of cream and make our own butter. Most of our cattle were either young beasts or cows suckling calves, and by February we were milking only two cows, Dorcas the Red Poll who was supposed to be due in April and was now nearly dry, and a very old Shorthorn, the Fossil, who had come to us at the drop in the previous August, and should still have been milking well had she not had an attack of red – water in October which had seriously lowered her yield. This is not a dairying country, and it is almost impossible to get heavy milking cows locally. But the risk of importing cattle from other districts, except as young calves, is very great. The driving rain and wind, the herbage of the acid soil deficient in lime, seems only to suit acclimatised beasts, while red-water fever, spread by the ticks which abound on the pastures, is specially dangerous to strange cattle. A good many years ago, when there were still large farms in Skye, the Laird bought a bunch of Galloway cows from Glenbrittle, one of the wettest and wildest parts of the island. They throve splendidly. But recently, when the number of Galloways had been reduced to nothing by the breaking up of large farms into smallholdings, he had to buy in fresh stock from the neighbourhood of Dingwall, and in the course of a few years he lost two-fifths of them with red-water or congestion of the stomach.

With this in mind I had insisted on a cow from the West Coast. But this did not save her; from the start the poor old Fossil was dogged by disaster. She was a good-looking beast, and had obviously been an excellent milker in her day. But as so often happens, she was sent to the sale too near her time, and when she calved here a few hours after arrival, her calf, a fine heifer one, was born dead. In spite of this set-back she milked well until the attack of red-water, when she began to fall off very badly, although we kept her indoors from January onwards and stuffed her with food. So that the question of another cow became serious. It must again be a West Coaster, and if possible young. The difficulty of getting a local cow guaranteed sound and quiet, with more than a pint or two of milk and *not old*, can hardly be realised until you try. I put an advertisement in the *North Star* for a young cow or heifer, calving February, West Coast essential. This brought a number of replies, most of them from completely inaccessible places, and all offering veterans of various degrees of antiquity. One wrote: 'I have a young cow twelve years old.' Another with but little command of English, remarked that her cow 'no care nor man nor woman is about her'. The most promising cow was at a place which though not very far away in miles, would have taken most of the day to reach, and I was just wondering what to do when a moon-faced lad came to the door announcing that his father, whom I knew by sight as a crofter in a township less than four miles distant, had a cow that might suit me. Tactful enquiries about the lady's age made her out to be eight; I mentally added on three or four years and concluded she might be worth seeing (as I could go on foot) before answering any of the letters. I told the boy that I would come over the following afternoon. About the same time I heard by roundabout gossip that the Laird's shepherd's brother had a good cow which he intended to send to the sale at Dingwall next day. He would be driving her over the hill to catch the evening train. There was no time to communicate with the shepherd's brother, but it was suggested that if I were to wait for him at a certain bridge six miles from Achnabo and one mile from the station, I could make (if so desired) my purchase, and he would bring the cow to me instead of taking her on to the train. The snag was that though I knew the time of the train, I had no idea when the shepherd's brother was likely to pass the bridge, as the trucking of a cow is generally the occasion of a prolonged pow-wow with everybody you know at or near the station; so that it was probable that the cow would start very early in order to leave time for all this, especially as she would be old and therefore a slow walker. The upshot of it all was that I didn't go to the bridge, and afterwards regretted that I had not, as the cow turned out to be a good cross Short horn, and most surprisingly, only six years old.

Next afternoon I walked over the hill to inspect the other cow. The moon-faced lad's father, a cheery Skyeman with a cast in his eye, met me at his byre door, and we spent some time politely circling round the object of my visit in ever-narrowing conversational spirals, until we got to the heart of the matter, and went in together to look at the cow. The stall was very dark, and I had her driven outside, where she skipped about in the deep mud of the courtyard with a nimbleness that did not altogether deceive the eye. She was a black-polled cow, more than eight years old, but large and well-shaped, and obviously correct in teats. I thought I might do worse; the man was short of keep and anxious to sell, and after a spell of bargaining a price was fixed and the cow delivered at the farm the following day.

For some days she moped, standing at the farthest corner of the field nearest to her old home. Then she settled down and began to eat. In spite of her owner's assurances, she did not look much like calving on 14 February. The Laird's wife, looking at her with expert and critical eye, pronounced that she would be a long time yet. We sighed, as we were getting too short of milk to run the separator. Meantime we racked our brains for a name. Her late owner had called her just The Black Cow, but as we had several other black cows, this would not do, and in the end we named her Jessie after the schoolmistress, whom she closely resembled. I think that Flora's sister, who was still at school, regarded this as not quite the thing. 'Katie', said Flora confidentially, 'will not believe that the name of Jessie Macfarlane is on the black cow.' In general, our names for the cattle – Priscilla, Dorcas, Brigid, Ruby, Pendeen, etc. – were unorthodox, and found but little favour with the natives. Alec, the first boy I employed, had a private set of names for his own use, while other people referred to the cows by some descriptive term – the Minister's Cow, the Blue Cow, the Old Cow, and so on. Our worst mistake in nomenclature was when Peter christened his collie pup Donas Beg, which is the Gaelic for 'Little Devil'. It was long before Flora would bring herself to call Donas anything but Puppy.

Time passed, and on the 13th Jessie looked less like calving than ever. But early on the morning of the 14th, when I went into the byre to milk the Fossil, who stood in the stall beside her, I saw that her shape had changed. She began to stamp restlessly. Peter had gone to feed the poultry, which at that time were housed at the bottom of the lower pasture. Spying Flora shaking a rug in front of the house, I gave her a shout, and she, ever alert where animals were concerned, dropped the rug and darted away to fetch Peter. But before they had time to return, Jessie, with masterly unconcern, had dropped a black heifer calf on the straw behind her. 'My name should be on that calf', said Flora in her solemn bass voice, when we had rubbed it dry and removed it to the pen; and it was called Flora.

The next job was to milk the cow. Having milked a good many in my time, I set to work without misgivings. The cow was quiet enough, but not one drop of milk could I extract. I laboured away, getting every moment hotter and crosser, but without success. Peter tried his manly grip – no result. Then Flora, but that was no better. The milk was there, but not one drop could we wring from her obstinate and rubbery teats. Meanwhile the calf, by now cold and hungry, set up a lamentable bawling, which we checked by ramming two raw eggs, shells and all, into her expectant mouth. Another attack on the cow met with no better fortune, so there was nothing for it but to call assistance. We sent for the wife of Rattray's shepherd, who had a great power with cattle. I had seen her charm the most refractory beast into submission, but I hated sending for her, partly because she was stone deaf, and also because she had a great contempt for my methods, and always read me a homily on the folly and unkindness of keeping cattle, especially calves, so much out of doors.

She came, walking at a great pace in gumboots. I explained the circumstances in a voice loud enough to be heard all over the glen. She understood, tied on an apron of sacking, shook her head, talked to Jessie in tones that suggested an immense pity for the evil fate that had sold her to so unkind and inexpert a mistress, and then sat down and began. The milk came in little spurts, then in a steady flow. Obviously Jessie liked her, but even so it must have been hard work, for the muscles of the woman's strong arms were bulging with effort. The calf was fed with something better than an egg, and the shepherd's wife, promising to come down again next morning if we could manage ourselves in the afternoon, went home, shaking her head over our boundless ineptitude.

We did manage in the afternoon, taking the job in turns; but the toughness of this cow was such that from this time until the middle of June, when no longer needing her milk for butter, we put a Guernsey calf on her to suck; we counted the business of milking Jessie the worst and most exhausting job of a strenuous day. It was then that we started calling her Aunt Dammit, and this is her name today.

The shepherd's wife came up again next morning, accompanied by her husband. At this time Duncan the shepherd daily visited some part of the farm, as Rattray had taken the wintering for his hoggs. He had the slow and gentle temper that is bred by a lifetime of walking on the hills and looking at sheep. But the Highlander, like other Celts, has queer fits of sudden anger, and I had seen Duncan himself in one of these furies. There was a croft attached to his house at the top of the glen in which he had his hay, corn, and potatoes. Having no horse, he used to get the land ploughed by the landlord. The first year I came to Achnabo, the whole estate was behind with the ploughing, and Duncan was told to ask me for a few drills for himself in my potato field. I gave him eight drills; he came and helped us to plant, bringing his own seed, but he got the benefit of my plough to open and close the drills, and of the potash fertiliser I was using in addition to dung. Later on he hoed his own drills, and was to have helped at the lifting; but the day I chose for this work clashed with a gathering of sheep, so that Duncan's nephew, a swarthy young man with a disingenuous eye, was sent to take his place. The potatoes were lifted, but the question was what to do with them. We could not spare the cart to take them up the glen; so it was arranged that they should be clamped at the side of the field, and removed later.

The potato ground, being a half-acre strip at the side of the corn, was not fenced. According to our usual practice, a grass mixture had been sown with the oats, and there was by this time a nice bite for the cattle. The potatoes were lifted, and all was clear for turning them in; but Duncan's two clamps were still there. So great is the passion of every class of stock for potatoes, that the clamps would have been torn open, rifled, and scattered in an hour. I sent up a polite message asking the shepherd to remove his potatoes as soon as convenient, and he would get my horse and cart to take them home. This was in the morning; by midday a fine drizzle had set in, and I was surprised to see Duncan and his wife working at one of the clamps. I went out to speak to them. They were grubbing away in grim silence; one clamp was nearly demolished and the contents put in sacks. I made some remark, which was answered by a grunt from Duncan, and a mutter from his wife about 'having things on other people's ground'. Something was badly wrong. As I stood there the drizzle became a downpour. It was silly to uncover potatoes in such weather; I told Duncan to get Dick and take the bags they had already filled into the shelter of the cart shed, and leave the second clamp for a better day. He growled something about getting a horse from the village, and made a feverish attack on the second clamp. I was really rather worried; he was obviously bent on sacrificing the potatoes to his bad temper, and I had no intention that he should. The wife was too deaf to hear, so I argued and pleaded with the man until he consented to leave the second clamp and take in the potatoes already bagged. The horse from the village was abandoned and Dick harnessed; and at the end of all this Duncan was himself again. A terrible to-do and all because he was asked to take home his potatoes! Next day the cattle went in. Deserting the young grass they made a bee-line for the potato strip, and began nosing about among the withered shaws on the chance of picking up a few tubers. But I had put the harrows over the ground, and there were not many left.

3

February: Some New Arrivals

THERE were two dwelling-houses at Achnabo – the farmhouse, and a cottage beside the bull-shed. This cottage, though intended for a ploughman or shepherd, was exactly like the farmhouse except that it was a little shorter and had no bathroom. Before Peter came my staff consisted of a girl friend who lived with me in the house, and a hired boy, Alec, who occupied a bedroom in the cottage, which was otherwise empty. By an arrangement that is common in Scotland for single farm servants, Alec had his meals in our kitchen, but was expected to clean his quarters in the cottage and make his bed. I used to go over once a week to change the sheets and retrieve various plates, bowls, and dishes which he had taken across with porridge for his collie and forgotten to return. The things to be seen in that cottage would have made a good housewife swoon with horror. Alec was a great letter-writer, and the floor, thick with the dust of ages, biscuit crumbs, silver paper, and nails, was liberally bespattered with ink. The rusty grate, in front of which was a stretched string full of damp decrepit stockings, was choked with ashes. The walls were hung with various articles of clothing, most of them in rags and all of them reeking of the byre. Downstairs, his litter had overflowed into the kitchen, which was full of newspapers, fish-hooks, cartridges, old boots, sticks, cracked cups, and rabbit skins in varying stages of decay. There are many lads who take a pride in keeping their quarters clean and tidy; and I think that Alec was perhaps unusually careless and messy. But except in lonely colonial farms, where such an arrangement is inevitable, the bothy system is a bad one. It takes an exceptional love of neatness and order to induce a boy, especially when alone, to keep his room and clothes in trim after a hard day's work. I did not blame Alec, but made up my mind to get some woman into the cottage, who could look after him and do some housework for us as well. Hitherto we had managed alone, but with the approach of summer and all its claims, I began to long for someone who could do a bit of cooking and cleaning.

I had the bright idea of writing to the local secretary of the British Legion in Inverness, to see if they could recommend a soldier's widow who would do part-time housework in return for a free cottage with milk, wood, and potatoes. After some correspondence I had a letter from what seemed to be a very suitable applicant. In my reply I told her all about the farm, not forgetting to stress its appalling loneliness. To this no objection was made, and in due course Mrs Clegg arrived, with all her worldly goods upon a lorry. She was a superior, capable-looking woman in her fifties, just what I wanted. With Alec's help she settled in her furniture, and arranged to come over at half past six every morning, light the stove, get breakfast, clean the house, and then return about half past twelve to cook dinner and wash-up after. This work was excellently done, and for a few days we were in clover. Then one evening Mrs Clegg came over in great distress, and collapsed in tears, saying that she must leave at once, as her nerves couldn't stand the awful solitude. She had not real-ised this desolation – how few neighbours, how small a farm, how far from the village. In the end I stopped her lamentations and persuaded her to stay for a month anyhow, seeing that she had brought all her furniture. After this outburst she settled down fairly well, except that she developed a fearful contempt for the farm, its tenants, and its ways. She was not a West Coast woman, and before marriage had been a housemaid in good service. It was obvious to her that a person who worked on her own farm and had her rooms so bare and cheaply furnished could not possibly be a lady, and her manners were made to fit this theory. Occasionally nerves would get the upper hand. One morning early, when she came to start the stove, I heard her muttering to herself and clanking ominously with the fire-irons. Something was very wrong. I hurried in, and found that we had forgotten to put kindlings on the rack to dry overnight. This of course is a grievous sin, but one that might be pardoned once or twice in people as busy out of doors as ourselves. I made polite apologies, and presented her with a bundle of dry sticks that I kept in reserve for emergencies. But even that did not appease her. 'What use are you?' she said bitterly. 'Always doing things that don't matter, and leaving undone the things that do. I can't think what God made you for.' I had often wondered that myself, but at this bleak hour the obvious retort escaped me, and I made no reply. The first duty of woman, and apparently the only one, was to put kindling to dry overnight. This new doctrine was rather disturbing – God moves in a mysterious way.

Five weeks later, when we had just begun to cut the hay, Mrs Clegg sold her furniture and went, leaving to me a few white plates and to Alec a broom with which he used to swish water over the dairy floor. The cottage once more became a bothy, and so remained until Alec's departure at the November term, 1933. As my girl friend, who had come for an experimental half-year, was leaving about the same time, I decided to try a new arrangement, with an educated boy of good class living in the house, and a maid in the kitchen. And so, when Peter and Flora were established, the cottage was abandoned to solitude.

But only for the time being. I did not want it to stand empty for long, so I induced the landlord to put in a bathroom, and then advertised it in *The Times*. I had over twenty replies, varying from needy parents who wished me to adopt their baby to deserted wives who wanted to forget their troubles in beautiful scenery. Of these only two were suitable, and after much correspondence the Gordons arrived, with several loads of handsome furniture. The cottage was not nearly ready; the bath had gone to the wrong station, the joiner was waiting for more wood, the plumber was waiting for the joiner to finish, and the mason who should have been excavating the septic tank was miles away at another job. Sometimes a workman would arrive about dinner-time on a motor cycle, smoke a few cigarettes, brew a cup of tea, make some hammering noises, and then go home. The weather was abominable. The furniture arrived in a snowstorm, and was somehow crammed into a cottage full of workmen's rubbish. The Gordons came to stay in our house, watching with impatient fury the protracted pageant of Highland *laissez-faire*.

The day of the Gordons' arrival was marked by another excitement. The bull came. I had gone to the station to meet them with the trap, and on our way up the last hill we passed Peter and three local boys and two dogs, with the bull on a halter, alternately towing and being towed. The Laird had bought him for me at a big sale of bulls at Inverness, and wired us to meet him at the station and handle him gently. I had arranged for Peter to have one companion, but this one, evidently expecting trouble, had brought two others. However, the bull was quiet enough, and when the procession reached the steading, the Laird was there to receive it and give detailed instruction about feeding and management.

We were very proud of George, but for the first week or two he was rather an anxiety. We lodged him in a spacious house adjoining the byre. The door of this house faced west and opened on to a passage about eight feet wide between the end of the byre block and the gable of the Gordons' cottage. The passage ran north and south, and in stormy weather it was almost impossible to stand there, especially when loaded with hay. George was fed with hay four times a day, and no small part of these meals was whirled up into the sky. The night after his arrival a gale sprang up, and by two in the morning it was blowing a hurricane from the south-west. As I lay awake it occurred to me that the door of the bull-house, which opened inwards and was fastened by a hook and chain attached to a staple, was not very secure, and that if the newcomer chose to get out he might be anywhere by morning. I pulled on my gumboots, kilted up my nightgown

28

under a mackintosh, put another waterproof over my head, and went out into the night. It was pouring with rain, but there must have been a moon somewhere, for it was not very dark. Once out of the garden the wind took charge, and I was whirled across the green to the bull-house passage. Here I could hardly keep a footing. The wind boomed against the buildings, slanting sideways on to the door of the bull-house, which was madly banging to and fro. As I had feared, the staple was working loose, so that every moment the door got more play. I remembered that there was a piece of rope in the byre. Groping about in the dark among the grunting cows, I found the rope, put a slip-knot round the handle of the door and made the end of the rope fast on the lee side of the building. My head covering had long since blown away, I knew not where; the rain was driving in long spears, and I was soaked to the skin. But George was secure. I fought my way back against the wind, stopping on the way to put extra stones on the roof of our new brooder house.

George had further to be groomed and exercised as well as fed and watered. Of water he consumed a great deal; pail after pail was drained, and when satisfied he would lick the face of the attendant with his rasping tongue, yawn elegantly, and turn away. He had six pounds of linseed cake daily. He was quite pleased to be groomed with a dandy brush on one side only, but the attempt to get round to the other side was met with a good deal of ponderous and rather alarming friskiness. Exercising was worse. He did not much care for the halter, and it was often difficult to put it on. Nor was he always ready to go in the direction indicated. Peter led him, while I marched beside with a stick. But one day we had such a tussle to get him back to the house that the exercise was omitted until it seemed safe to put him out unhaltered with the Fossil and Aunt Dammit. So soothing was the influence of feminine society that henceforward he was always out with the cows, and there was no quieter beast on the place than sleek, complacent George with his doting harem.

In a country of smallholders, the owner of a good bull can

make large profits. Fees very from 7s 6d to 12s 6d, and if the bull is given plenty of cake he can serve sixty or more cows in the season. For crofters' cows, a pure-bred Aberdeen Angus is the best. Townships of a certain size have a bull supplied to them by the Department of Agriculture, but the smaller groups have to depend on privately owned animals, and if the official bulls are for any reason not approved of the crofters will go to a private owner. Luckily for us, one of the neighbouring government bulls was a Shorthorn, while the other, though black, was very small and had the reputation of leaving undersized calves. So that we got a good deal of extra custom.

The previous season I had subsisted on borrowed bulls. The first one, lent by the landlord, had been a fine animal in his day, but had grown fat and sluggish with over-feeding, and took all day to serve a cow. So that people who had anything much to do with their time gave up coming. One old man, who had spent the whole afternoon watching a protracted and ineffectual courtship, apologised for staying so long. 'One right jump', he said, 'and I'll take her home.' By evening he had gone; and as I found three half-crowns laid on a milking stool in the byre, I concluded that he had got his 'right jump'.

Such slowness was too exasperating to endure for long, so I returned the sleepy veteran and took a loan of the Laird's White Shorthorn, of whom I have more to say later. By this time he had served all the cows at Strathascaig, and as he was a practised gate-crasher, the Laird was only too thankful to get rid of him. At Achnabo he broke one gate, but otherwise was little trouble because we kept him in a field with stone dykes; and as soon as our cows were all served he was returned and sold.

The only troublesome bull I have known was Hannibal, a young Guernsey reared by the Laird and kept for a year or two to get some pure-bred Guernsey calves. A field, fenced with extra high courses of barbed wire, was set apart for him. On the river side of this field, just inside the fence, was a bank, made long ago for protection from flood. The top of this bank was worn bare with continual pacing. To and fro, up and down, often for hours at a stretch, Hannibal would pace, bellowing horribly; some wag called this performance the promenade concert. At intervals he would stop to paw the ground or ram his head into the bank, plastering his face with reddish mud. In more lively moments he would smash his drinking-tub to little bits, or uproot a few young trees. The sight of sheets in a gale volleying and thundering on the clothes-line, or of anyone with plough or harrows, would drive him into a frenzy of irritation. I cannot say that he ever actually chased anyone, but he would follow people down the fence with no very friendly gestures, stopping now and then to paw the ground and roar. I remember once crossing his field to repair a gate on the farther side. I had a stick, but a powerful blow on the nose seemed to worry him as little as the alighting of a fly on one's own. I walked slowly, not daring to hurry, wagging my stick behind me like a tail. He followed, moaning a little, quite close, so that I seemed to feel him blowing on my neck. If I turned, he would execute a little war-dance with lowered head. This was all that ever happened; but he was still young, and one never knows. Of course he had his quiet moments. Standing in his stall in winter, he would turn to suck my fingers as if still a baby; and on Christmas Eve he let me put a paper garland round his neck and tie sprigs of holly to his horns. He had a passion for treacle, and would suck the neck of an empty treacle bottle with all heaven in his eyes. But we took no risks. When he had left enough calves, the Laird sold him, and he was last seen walking down to the station in proud isolation at the head of a drove of miscellaneous cattle. Poor Hannibal! It grieves me to think of him slain to make soup for the smug inhabitants of Dingwall. He must have been too tough for anything else. He is now a legend, a kind of bogy-man, like Boney.

His successor was Abdul the Bulbul Ameer, a white Shorthorn, quieter to handle than a cow, but of an incurably roving disposition. Being specially selected for crossing with the Galloways, he spent the summer months in their company; and as long as no other cows were in sight, would sit at peace with his dusky harem, like a complacent eastern trader, chewing his cud. But let some new female, hoary veteran, or even calf a couple of months old, be seen passing along the road, and off he would go, over gates and through fences, to seek her society. No barrier could retain him. He had a trick of going down on his knees and lifting the wires with his head, until he had made a gap large enough to wriggle through. This wandering habit made him more dreaded than he need have been, considering the mildness of his nature, for he was always appearing in unexpected and unsuitable places. It was disconcerting to go into the henhouse, and find there the muddy hindquarters of the bull; or to see a gleam of white among the trees where the clothes hung and discover that it was not a sheet but the bull. At this point, however, let me say that bull-fearing people are terribly self-centred. They imagine that the poor brute has no object in life but to annoy them. But apart from food, sex is his main concern, and since, being human, they cannot very well arouse the bull's jealousy, they do not come into conflict with his strongest instincts, and therefore do not seriously provoke his anger. A cow with a calf is much more dangerous to approach, because the presence of a stranger rouses the protective instinct of a mother. I am not of course suggesting that one should take unnecessary risks with bulls, especially with strange ones; but a good deal of the terror they inspire is quite unreasonable. A bull at large, with plenty of female society, is a peaceful and contented being, lacking nothing but the spice of rivalry. He is most unlikely to annoy anybody. Two anglers returning from their fishing saw a bull grazing beside an Irish cabin. They hesitated, and were about to give him a wide berth when an old woman ran out, slapped the bull with her open palm, and said: 'Get along wid ye! Would you be after frightening the gintlemen?' They felt rather small.

There was something pleasing about the whiteness of Abdul. His colour, the symmetrical spread of his beautiful horns, and the enchanting curls on his forehead, reminded me of the sacrificial bulls we read about in the classics, and I longed to put a garland of roses round his neck and lead him solemnly round the fields in the revival of some ancient rural ceremony. And there was use as well as beauty in his whiteness. In the daytime, a white bull can be spotted miles away; and even in the darkest night his colour is faintly visible, so that you cannot stumble on him unawares. An all-black beast seems to focus and intensify the surrounding gloom; a dark beast with white markings shows only the white, a queer apparition. One pit-mirk night I was walking beside the burn, guiding my steps by the sound of the water, when I saw suspended in the air what looked like a luminous dagger. I put out my hand, and touched the white-starred forehead of the horse. The Galloways were not far off, but invisible. Had I not heard them breathing, I should have fallen over them.

To own a bull is, as I have said, a profitable business, but it is something of a tie. There are no fixed hours or Sundays off: cows sometimes arrived soon after six in the morning or long after eleven at night, and on more than one occasion Peter, the official bull-herd, had to be dug out from bed or bath to attend to some importunate client. One evening when no one was at home but a girl who was helping with the hay, I had gone to bed early after a strenuous day, and was just drifting into a doze when I heard a knock at the door. The girl answered, and there followed a long confabulation. I was beginning to wonder what it was all about, when a bearded face appeared at the window of my ground-floor bedroom. 'You are sleeping,' a voice remarked. When I said No, it continued: 'Will I get the bull for my cow?' 'Yes,' I replied, 'but you must take her down yourself. The bull is in the park below the barn.' Off he went, and once more I turned to slumber. After a while the apparition returned, enquiring the price; and hearing it, laid the customary three half-crowns on the windowsill and vanished.

We always sent someone to administer the bull, and if Peter was out Mr Gordon would volunteer to go, for local propriety did not approve of a woman giving the bull except to a customer of her own sex. We found women clients rather a nuisance, because they often brought cows that were not bulling, and were then annoyed because they did not get served. Also, they were usually too nervous to take the cow into the field themselves. One evening in early August, when the nights were beginning to get properly dark, a woman came to the door in great agitation. She had brought a cow for the bull, and had been walking the world for several hours. I had intended to send her away, since it was late and dark and Peter was in bed. But she began to explain tearfully and at great length that she was a lone widow and had no man of her own to send with her cow, and could not afford to pay anyone, and was so tired with walking and tomorrow would be too late. So I relented and took her off in search of the bull, whom I believed to be somewhere in the lower pasture with the Belties. By this time it was too dark to distinguish any black beast, but after tacking up and down the field with the widow and her cow in tow, I spied at last something white, which proved to be the belted part of one of the Galloways lying down, and near her was George, totally black and lost in darkness, except where gleams of vanishing day caught the ring on his nose. I poked him up: he was sleepy and disinclined to be amorous. The widow was horribly agitated, torn between fear of the bull and the desire to get her cow served quickly. At last, with much coaxing, the deed was done, and pressing some sticky coins into my hand she set forth upon the long trek home. She never came back, so I assume that the cow proved in calf.

It is necessary, as I have said, for someone to administer the bull; for the crofters, if they were allowed, would try to get several consecutive services 'to make sure', and though most farmers insist on the fee being brought with the cow, many people would try to postpone payment till they were certain she was in calf, unless somebody from the farm was there to wring it from them. It was also darkly hinted that certain people put in their cows by night without coming to the farm at all. 'I may know something or I may not,' said one of my crofter friends in a meaning tone.

'Of course you can please yourself, but if I were you I should put a lock on that gate.' I took his advice.

Before we had a bull on the premises, much time was wasted in driving cows to distant and expensive mates. There was one Cornish Guernsey who needed three or four services to get in calf: Alec used to assert that he had worn out his shoes taking Pendeen to the bull. Poor Pendeen! She has now paid the price of her anti-social practices. Having visited George with unfailing regularity from February to October, and all without result, we grudged her winter keep and sent her to Dingwall, where she realised the magnificent price of £1.

Soon after we went to Achnabo our landlord bought a young Dairy Shorthorn bull with a portentous milk pedigree, in order to provide the crofters with heifer calves of good milking strain. Having at the time a Guernsey cross-bred heifer ready for service, I decided to send her over, and arranged with the farm manager to keep her for a few days, and let us know when she was served. I did not want to pay some boy or man a day's wages to drive her there, so Peter and I planned to go ourselves. It was early in January, and having our housework to do first, we did not start till about half past eleven, which should have given us plenty of time to get home before dark. Banking up the range with small coal and putting kindlings to dry, we set out, driving in front of us the heifer and two other beasts for company. All went well till we reached the main road, where we ran into a large mob of cattle which the crofter women were driving out to pasture. In winter they used to keep the cows in till noon, feeding them on corn and hay, of which they always seemed to have an abundance. We should have remembered this and started earlier. A free fight began; the hillside resounded with yells in Gaelic, the barking of collies, and the thud of sticks on callous hides. At last we hustled our beasts through and resumed our journey. The morning had been fine, but now dark scuds of cloud were flying along, driven by a gusty and ever-increasing wind from the south. We were ascending the northern or leeward side of the pass, so that we did not meet the full force of the gale till we reached the top, when we had to fight for every inch of progress. The hills of Skye and Glenelg were already blotted out by swiftly moving curtains of rain, which soon enveloped us in a horizontal downpour. The cattle put down their heads, and, urged with sticks, plodded steadily on. By the time we reached the farm we were wet through, and decided to get some tea at the hotel which was about a mile farther on. Leaving our beasts with a boy at the farm, we dripped into the deserted dining-room of the hotel, and ate the largest and best tea I had ever tasted. We could not afford much time, as darkness would soon be falling, and there was no one left at home to keep the fire going or take in the cows. The shortest way from the hotel back to the farm was to cut across the sand and shingle of the little bay. Here we caught the full force of the storm, being broadside on to the wind, whose violence was intensified by being sucked up the narrow funnel of the glen. The bay was cut in two by a stream. Too wet to bother about going round by the bridge, we plunged in. The bottom was of shifting sand and pebbles, the water deep, swift, and icy cold. In the middle it reached half-way up the thigh. We got across, emptied out our shoes, picked up the companion cattle at the farm, and started home. Mercifully we now had the wind and rain at our backs. We reached the top, and were swept down the other side at a run, the cows' tails streaming before them, and the dog's coat parted to the skin by vicious flaws of wind. Trees groaned and shrieked; the rain swished down without mercy, the long road still before us glimmered in the gathering darkness with puddles full of ruffled water. Reaching home, we could just make out a bunch of cattle huddled at the gate, tails to the weather, dumbly enduring. We let them in and then went to the house. The fire was out, and in the gloom we groped for matches and the kindlings we had set to dry. Passing my hand over the top of the stove I found not sticks but a little heap of warm ashes!

After the Gordons and the bull came the snow, which is one of the terrors. A terror is not just something that frightens: a bomb or a rat could do that. It is a thing of beauty and power – beauty inhuman and mysterious, power almost intolerable, which awes us into silence, and compels the soul to worship and abasement. Snow is one of the terrors. For your peace of mind, it is best to treat it as a mere nuisance, blocking the roads, clogging your shoes, treading over the house. Or to think of lambs buried in April snowstorms, men lost in drifts not half a mile from their own door, or even of Antarctic blizzards or Himalayan avalanches. These aspects of snow are alarming, but somehow within human scope, to be fought and guarded against. Most of us have the sense to defend ourselves from the intolerable magnitudes around us by concentrating upon practical utility. We use the stars to steer our ships, and electric power to light our houses. To most of us, electricity suggests not lightning or Aurora Borealis, but a set of handy gadgets, and the stars a pleasant Saturday evening at the pictures. We need not damn our limited vision; it serves as a kind of smoked glass to save us from blindness. For who can see God and live?

Once think of snow in an unpractical fashion and it becomes a terror. Its whiteness, its coldness, its stealthy silence, combine to appal us. Of whiteness Melville says: 'There lurks an elusive something in the innermost idea of this hue which strikes more of panic to the soul than the redness which affrights in blood.' Whiter than snow is a fearful comparison, since snow will make even blown spray or the purest bird's plumage look dirty. The harsh light cast up from the ground or filtered down through a snow-covered skylight turns the ruddiest face to the ghastly resemblance of a corpse. In these days, when it appears that the end of life on this planet will come through loss of heat, cold is the truest symbol of death, and snow and ice more dreadful than the flames of the fiery furnace. The silence of snow, like that of fog, is deceptive and dangerous. Lightning and flood destroy with warning uproar; but snow works in quiet and secrecy, drifting, enfolding, obliterating. It kills in sleep, with soft imperceptible caress, though no one has come back to tell us what pain there may be to suffer before the end.

On the West Coast, snow does not lie long except on the higher hills and those farthest from the sea. Nor does it come early; there is often more to be seen in June than in December. This year it arrived very suddenly. A mild wet spell ended with a shift of wind to the north-west, bringing first sleet and finally snow. It was blowing a gale, and the snow came in heavy showers of increasing force and length with brief intervals of struggling sunshine. During the showers the air was thick with blizzard, the world empty of form and colour. From time to time part of the horizon would clear, revealing smoke-like whirls of snow and vapour, dark in the air, but whitish against the purple of the hills. Rifts in the clouds showed cold blue strips of sky, fringed with the brilliant mare's-tails that mark the edge of a storm. A fresh snow-cloud, swaying and eddying like a blown curtain, advanced across the sea, the iron-grey water turned to black before it, and then, as the squall struck it, broke into snarling foam. A ray of vanishing sunlight caught the cloud, giving it for one moment the look of fiery smoke: then all was swallowed up in whirling gloom. The ground was quickly covered with sharp many-faceted pellets which cut the face without mercy. The shaggy Galloways, lined up to get their daily feed of corn, were soon frosted over like Christmas cakes. Beside the burn, the shepherd was fighting the gale with a big bundle of sheaves in his arms. The hungry cows began to mob him, pulling out whatever loose stalks the wind had spared. Little by little he forced his way to the shelter of the whins and threw down the bundle, making a fantastic splash of colour in a world of white: green spines with their yellow blooms, black cattle, the tawny gold of the corn and the warm crotal of the man's coat. All this I watched from the window, thanking Heaven that it was not quite time for me to go out.

I remember one evening at Strathascaig, about an hour after sunset. Winter twilight is short: the arch of luminous sky is narrowed by the hills, so that the glen becomes a cup of premature darkness. But the snow had changed all that. The whole earth glimmered with a stark and pallid radiance; tracts of stratified black showed where the smoother contours of the hills were broken by cliffs of naked rock. Under the whins, where no snow lay, was inky gloom; a Galloway, invisible in her blackness, only betrayed her presence by a long snoring breath. Apart from these scattered and as it were accidental patches of dark, the earth was lighter than the sky. That night I saw a most uncanny moonrise. The east was netted and festooned with finest threads of cirrus. The moon, now in its last quarter, sprawled supine upon the pale rim of the hills. Slowly it rose, an uncouth semi-circle, barred and distorted with vapour; and before it could heave itself clear of the mists, a small black cloud shaped like a deerhound's head, leaped up from the horizon and raced across its blurred and sickly face.

In three or four days the snow had disappeared and the steading was once more a sea of mud. While it lasted we gave the byre cattle extra rations, but to the scandal of our neighbours we put them out at nine o'clock as usual, driving them to a field full of rough bent-grass and rushes, which projected from the snowy covering and gave them something to pick at. There were far too many rushes at Achnabo, but one or two rushy fields are of great value to cattle in winter and early spring, when the young shoots, with the coarse grass that grows round them, resist the weather and provide a considerable amount of grazing. The Galloways, safe in the shelter of the woods, were out day and night, on a ration of three oat-sheaves a head. Some young beasts, from nine to ten months old, spent the day in one of the lower fields, in which was a little spinney of birch and hazel. From this place of refuge we had some difficulty in extricating them at night, for a fierce wind was lifting the dry powdery snow from the ground and sweeping it across the field like spindrift, in sharp blinding volleys they did not care to face.

At this time we rather regretted having banished the hens to the lower pasture, as it meant quite a long walk three times a day. Before the coming of Peter, who had worked on a scientific poultry farm, the hens were rather a joke. Some were a mixed lot bought from Strathascaig, the rest a few nerve-racked survivors from a disastrous brooder fire in the spring, most of whom succumbed later to fowl paralysis. Not one of these birds would settle in the house intended for them. They roosted on gates, on the tops of old boxes, on bags in the cart-shed, on the rafters of the sawmill. They scratched about the steading, laying eggs in all sorts of odd places, spoiling the hay, the corn, the garden, the granary. So Peter built a handsome henhouse out of old timber, and thatched it with rushes. We carted it down to the lower pasture, fixed up a run of wire netting, made a nocturnal raid on the roosting ones, crammed them into bags and conveyed them, squawking and struggling, to their new home.

Peter was always pessimistic about his luck with poultry. A few nights later his forebodings were justified. Some beast, probably a wildcat, raided the hen house, killed two Rhode Island pullets, mauled another, and gave the rest such a fright that no eggs were laid for a fortnight. A heavy trap was set, but without result. These cats, which are now becoming very rare even in Scotland, are not infrequently seen in this district, and they are responsible for the destruction of a number of hens every winter. Nobody who has seen one could ever take it for a domestic cat gone wild. The tawny coat with its handsome markings, the short bushy tail, above all the savage head with baleful eyes and snarling lips, leave the impression of something rare and distinctive. Once when driving in the trap quite close to the farm, we heard Thos barking at the foot of a young tree not twenty yards from the road. A wildcat dashed up the trunk and crouched, swaying on the topmost branch, glaring at the dog with a look of indescribable ferocity. Peter dashed for a gun, but before he could get back the cat had disappeared.

After a time we removed the run and let the hens roam at large. Little by little they worked their way upward, until the day came when they were all back in the steading. Peter then made a portable fold, from which it was impossible for the most athletic or discontented bird to escape, and after that there was peace.

We then borrowed the Laird's incubator and set a collection of eggs sent from Peter's home. For some unknown reason only a dozen or so hatched. We made up the number with Light Sussex day-old chicks, and put them into an ingenious brooder which Peter had built. The weather was abominable, and the little wretches pined and died daily. The food and temperature were all correct; but I believe that successful chicken-rearing is largely a matter of luck, as the Laird reared 97 out of 100 on precisely the same management as ours. We had to return the incubator, so it became necessary to buy one of our own. It was a second-hand one, and must have had some devil in it, as only one out of a hundred eggs hatched! Undiscouraged, we bought more day-old chicks, cross-bred ones this time, and a ready-made brooder. For a time all went well; then one day black smoke was seen issuing from the brooder. Water was rushed from the well; the outer house was saved, but the inner was burnt out and a few weakly chicks incinerated. Others were drenched with water and had to be warmed back to life in the kitchen. The whole fraternity got a bad fright and the death-roll began to increase. Such chickens as survived cold, fire, and the ravages of hoodies grew up fine and hardy; but it seems to me that the person who undertakes to rear chicks must have an exceptional stock of patience, faith, hope, and charity.

4

Interlude: A Winter Night

EIGHT o'clock, and a dark winter's night. On the hearth a block of green birch sizzled dreamily – soft lamplight fell on open books. Now and again a gust swooped down the chimney and stirred the embers, assailing us with puffs of fragrant smoke. We sighed, looked at the clock, and left the social blaze, cramming reluctant feet into damp gumboots that seemed many sizes too small. Milk pails and lanterns were collected from the kitchen; a mass of scrambling, barking dogs shot through the open door and we followed them into the night. Outside, the mild air was thick with drizzle and full of the noise of a south wind that had been blowing for several days. The sound of the falls on the river, increasing and diminishing with every fitful gust, was like the ebb and flow of waves upon a not very distant shore. Nearer and more continuous was the moaning of the wind in the pines that circled the home fields. The light on Crowlin winked mistily across the Sound. We entered the byre and hung the lantern on a nail; its beams fell on a row of bony rumps and switching tails. Every stall was occupied, and the cows' breath kept the place warm and comfortable. As we entered, the cattle craned their heads over the high partitions, like people trying to catch a glimpse of royalty. There was a big heap of hay in one corner, and we started to fork it into the racks. This was the signal for a veritable war-dance. Chains rattled, feet stamped, tongues flickered, eyes goggled, necks were stretched giraffe-like as far as they would go. The animal-lover must often be distressed by his charges' lack of faith, or even of observation, since day after day the last beast to be served abandons hope of getting its portion without thieving from a neighbour. Tossing their heads, they rent the bulky stuff, picking the choicest morsels. As I served them, I wondered where this particular lot was cut, whether it was easily or hardly saved, and whether it was myself, Peter, or Vivian who pitched it into its final place in the sawmill. Who knows? Only some day, perhaps about Candlemas, I shall hope to find a woollen cardigan, stripped in the heat of stacking, which someone forked up in the hay; and there it must remain till we have fed down to its level.

Having fed the cows, we sat down to milk them. My own stool was home-made, with legs curiously straddled, because I was too impatient to find the correct angle for boring the holes. It was rather like a decrepit mediaeval war-horse, but this did not matter much, for the irregularity of the legs seemed in some way to suit the slope of the floor. The spurting jets of milk, forced into an empty pail, made loud metallic noises, which gradually diminished as the vessel filled, until, when the frothing surface neared the top, they faded into a rich purring murmur. How many times since the beginning of human life, in all the byres in the world, has this ancient and soothing music been heard, inevitable as sunrise and sunset, soft as a lullaby, healing sorrow, promising peace! To appreciate all this, the reader must of course assume that the cow stands quietly, and does not put her foot in the pail or send stool and milker flying to the other end of the cow-shed. Otherwise the divine peace of milking-time will degenerate into a profane imitation of a circus. Our cows were fortunately stolid, and we could lean our heads indolently against their flanks and meditate or sing. When milking together, especially if some visitor was in occupation of the sitting-room, we would prolong the milking by singing hymns, which cannot have sounded very pleasant, because I had not much voice and Peter could only sing in semitones.

The first cows I ever milked were in the old cobbled byre at Strathascaig. It was long since the Laird's cattle had enjoyed the attentions of anyone but their master. The last regular 'man' was

Jimmy, an ex-fisherman from Buckie, who had spent most of his life in trawlers, and had but little aptitude for farming. He used to work in an ancient blue jersey full of holes, and had immensely long bushy eyebrows which he could twiddle like the antennae of an insect. He may have been a hero afloat, but was horribly afraid of the bull; and I remember how he once rushed into the house to find the Laird, exclaiming: 'The cows are fast, but the bull's adrift!'

It was always difficult to get a maid who would milk the cows, and for some time the Laird employed Morag, the shepherd's housekeeper, an old woman from Skye, who had a wonderful power with cattle, but her knowledge of English did not extend beyond a few well-worn phrases, and to venture further in conversation was a very risky business.

One summer evening I ventured boldly: 'The midges are very bad tonight, Morag'. To which she replied: 'Beautiful indeed!' Another time I was even more enterprising. 'I hear that Archie's Donald has the measles.' 'Better for him indeed!' was the uncompromising answer. After that I stuck to safe platitudes. Having agreed that it was a wet night, or a fine one, as the case might be, Morag turned the stream of her discourse upon the cows. 'Ach, Maire, the beautiful cow. Are you here then, Maire? Are you pleased tonight? Are you getting hay, Maire?' Then suddenly the cow lifted an impatient foot, and the crooning monotone gave way to a stream of angry noises, in what language I could not say. It was not English, and I do not think it was Gaelic. It may have been some secret tongue understood only by cattle, for Morag had an uncanny knowledge of beasts.

The shepherd, a widower with three sons at home, kept her as housekeeper, and she served him faithfully, though he complained that she was too fat to bend over the girdle and turn his scones, so that they were often burnt to a cinder. Fat she was, indeed: so fat that I often wondered how she ever managed to sit to a cow, or cross the long footbridge without going through. Her girth was increased by many layers of thick woollen clothes. She would roll down from the shepherd's like a ship in full sail, having first cleaned and preened as if on the way to attend a levee. Yet in spite of all cleaning and preening, I doubt if our pure milk pundits would have allowed her near a byre. When she came with her pails to the dairy door, where (better for them indeed) they were always handled by someone else, there was wafted into the house a subtle and penetrating odour of cow. But what about it? Better the aroma of a thousand cows than the stink of an exhaust.

One day she brought us a letter from a cousin with whom, when not in a situation, she used to make her home. He pointed out that he could no longer look after his bedridden mother alone; Morag must leave the shepherd's service at once, and come back to Skye, otherwise she need not expect to find a home there when she wanted it. The writer's motive came out in a postscript: 'You are old enough to get the pension next month, and you had better see about it.' She did not want to go, but in the end she went. I think she regretted the cows more than the shepherd.

About a year later I went to Skye to fetch home a pony, and found that I should have to pass through the township where Morag's cousin had his croft. After a ride of about twelve miles I reached the place, and then suddenly remembered that I knew neither Morag's surname nor that of her cousin. Seeing a man on the road, I asked him if he knew where a very fat woman lived, who was called Morag, and had come from the mainland to stay with her cousin. He thought for a moment, and then said that I must mean Morag Macintyre, who lived in the farthest croft up the brae. He waved his arm towards an apparently empty hillside. I put my pony up the brae, and rode over bogs and rocks until I reached a dilapidated thatched cottage. Seeing no one, I did not dismount, but let out a few shouts, which brought an old woman to the door, very dirty and almost without English. I asked for Morag Macintyre. She took my bridle, and led me across more bogs, until we met another crone, even dirtier than the first. She beamed and shook my hand with fervour. It may have been Morag Macintyre, but it certainly was not my Morag. So little English had the pair that it took me ten minutes to explain that this was not the Morag I was looking for; that I was quite sure of this, because the Morag I wanted was well known to me, and I could *see* that Morag Macintyre was not she. At last with much regret they let me go. Farther along the road I found a second man, to whom I put the same question. He directed me to another house, which was hidden behind a clump of bushes. This time I was lucky. An old woman came to the door, but it was Morag, my own Morag. She was very pleased to see me, but not at all surprised: I might have been standing at the shepherd's door at home. After so long a search I could not spare more than a few minutes, but she had time to ask after all the cows. I don't suppose I shall ever see her again. Dear Morag! Dark woollen cloth, peat-soot and grease and aroma of cow. I don't much like any of them, but under this vesture of decay was a character, one who could neither read nor write, but was more amusing company than many a learned wit. Better for us indeed!

Morag was succeeded by Mrs Macfadyen. This lady came in reply to the Laird's advertisement for Trustworthy Woman (child no objection) to milk a few cows, cottage provided. The child turned out to be her granddaughter, and the woman, who called herself fifty-eight, looked ten years older, and so she proved to be. In a moment of domestic crisis and blind recklessness she was taken into the house to cook and clean. She cooked in a savoury but greasy fashion, and cleaned neither herself nor her kitchen. With this I was not concerned; I knew her only as a milkmaid, and if Morag was a strange apparition, Mrs Macfadyen was an even stranger. She would come into the byre wrapped in a ragged and incredibly dirty overall, her head turbaned with bands of filthy white and scarlet silk wound round a kind of skull cap of brown paper, flapping and trailing along in carpet slippers to protect her corns from the cobblestones. In one hand she carried a branch of spruce as tall as herself, and in the other a quart measure. She had once been dairymaid to a well-known Highland proprietor; but either this high-born chieftain had little care for the state of his dairy, or else Mrs Macfadyen did not think a plain farm establishment worth washing for, as some people will not dress to dine off cold mutton. For Strathascaig ways she had a supreme contempt. 'What a place!' I once heard her say to her grandchild. 'Cows and calves all mixed up like the dogs' dinner!'

But next to her dirt, the most remarkable thing about Mrs Macfadyen was the number of things she was afraid of. Wind, rain, burns, bridges, cocks, ganders, bulls, and newly calved heifers were all terrible to her. She was one of those people who must always be in the limelight, the long-suffering victim of unsympathetic treatment and ailments innumerable. These complaints I would meet either with stolid silence, or with some ruthless comment in the style of Mr Harry Graham. She had the early Victorian delusion that the more timid and helpless a woman is the more attractive she must be. A newly calved heifer moved her feet and tail in rather a restless fashion. At once the old dame roped her like a homicidal maniac, and for the rest of the season milked her into a measure held in the hand, standing all the while in a stooping position – a most wearisome business, and very trying for her aches and pains. But the bull moved her to a perfect frenzy of terror. If she saw his white bulk looming at the farther end of the field, she would kilt up her skirts, and, corns or no corns, scuttle to her cottage with the speed of a Wimbledon tennis champion.

To return to ourselves. We finished the milking, shut the door upon the cows serenely munching hay, and went out into the night. The wind had veered west, the air was much colder, the sky clear. There was no longer darkness, but a diffused radiance, something more than mere starlight – a radiance like that which is seen in the east when the rising moon is just below the horizon. Open stretches of road and the water in puddles glimmered with light from no visible source, as if the dome of the sky were made of some faintly pervious material, transmitting the distant brightness of someone else's day. The cause of this radiance seems a mystery. It is most often seen at seasons when the sun is very far below the horizon. It may have some connection with the displays of aurora which are fairly common here in winter. But whatever its origin, it is one of the most intriguing and beautiful things in Nature.

We returned to our fireside and our books. This was the best month of the twelve and the best hour of the twenty-four. The year's work was done, as also the task of the day; and a vast satisfaction in things accomplished, such as I felt renewed each time I contemplated the sawmill and the barn packed with winter keep, spread deliciously through every fibre of my being. Now for a good book. I am sure that hand work and brain work should never be separated. Taken singly they spoil a man's balance; but when they are combined in due proportion, not only is the work better done, but it is more enjoyed. Modern specialism is, of course, all against this. Hodge must hoe turnips (or whatever corresponds seasonally to turnip-hoeing) all day and every day, while Dryasdust must sit from morning till night with his nose in a book. It is often thought that manual work kills mental energy, and makes a man too torpid or merely too tired to use his brains. This I find from experience is a mistake, or at least is only true of excessive physical labour. Provided that there is no overstrain, the system is best rested not by inertia but by a change of occupation. To be properly appreciated, reading should be done at night before the fire, after a day of outdoor work. When as a university student I spent whole mornings at my books, I took but little real pleasure in them, certainly nothing to be compared with the joy of a volume saved till evening. A Quaker lady of my acquaintance used to disapprove of reading in the morning. I often wondered why, but now I know. He who reads in the morning will never taste the delights of an hour with a book, snatched from the jaws of a thousand practical claims.

Somewhere in *Moby Dick* – I would rather quote than paraphrase, but have searched the book a dozen times in vain – somewhere in *Moby Dick* we read that it is the adventures of the spirit rather than the exploits of the body that move and fascinate the young. At first sight this seems surprising: for adventure is the heritage of youth, and by adventure we mostly mean not contemplation, but action. To live like Captain Ahab is adventurous, but not to live like Kant: to discover a new continent, but not to invent a new system of thought. Yes; but very young people, having been but recently made free of the world of culture, cannot help being a little priggish, even a little snobbish, about it. A short time ago, Marryat and Stevenson held them spellbound; but they have put away childish things, and physical adventure leaves them cold. They no longer want to be sailors, explorers, engine-drivers: all that is as dead and despicable as model trains and catapults. Now they read poetry and philosophy; what wonder if they want to be poets and philosophers, and share in the spiritual adventures proper to the life of contemplation. If I no longer desire to live Miltonically and love Platonically, if I delight in the exploits of brave men in tight places, if I feel that the intellectual adventures to be met with in academic groves are a little nebulous and inadequate, it is because I am getting older, and have lost power and imagination, so that I need to be stimulated by outwardly exciting events. It makes one ashamed, like admitting that one's religious devotion cannot be roused by the silence of a Quaker meeting, but only by the blaring trumpets and drums of the Salvation Army. The course of life must be curving back to that pre-cultural paradise of Indians, clipper ships, and pirates, and I myself becoming like the regressive parent who buys toys for his children in order to play with them himself.

When Melville wrote this passage, he was thirty-two, a fully developed man in the prime of his power. The hard exacting life in sail is quick to turn boys into men, and at this age he must have had a far maturer wisdom than could be dreamed of by a professor of fifty. He was no longer young; he had passed through the green sickness of speculation, or he could not have written as he did. He had also passed through the reaction or 'rebarbarisation' stage, when outraged nature tries to rectify our youthful bias towards naked and rootless thought. He was within sight of that ultimate wisdom so often delayed till we are too old to make use of it, when flesh and spirit, action and contemplation, poetry and common sense are no longer opposed, but parts of one harmonious consciousness. The narrow, specialised philosophy of youth develops into the wisdom of the masthead, where the watcher, slung between the immensities of sea and sky, is free to contemplate God and the universe, provided that he also keeps his weather eye open and sings out every time.

It is generally thought that no one who is too young to remember the Crimean War should write an autobiography, because, unless you have seen a great many things and met a great many people, your motive for writing must be pure self-glorification. It is true that very old people have an immense lot to remember; but on the other hand they often remember it very badly, so that their reminiscences become a mere string of tedious anecdotes. Nor need they be any less self-centred than the young, since old unhappy far-off things are often recorded not for their own interest but because the writer took part in them. I admit that the autobiography of a very young man sometimes gives me the feeling that he regards himself and his doings as 'news'. So of course they are; nothing can ever be so vivid to us as our own lives, so that it is comparatively easy for us to make them interesting to others. We are naturally, even necessarily, self-centred; all literature, even the most impersonal, is at bottom autobiographical. But even conscious autobiography need not always be mere self-display. A man may write his life to serve some altruistic end, to show for instance how salvation may be won through correct breathing, or damnation risked by drinking too much tea. The few autobiographical details that follow may be interpreted as the reader pleases. They may be taken as a tract against the higher education of women, or a warning against letting university graduates dabble in the morning dew; as a sermon to parents on the desirability of marrying off their daughters, or on the ill effects of allowing them to spend the spring season on a farm. You pay your money and you take your choice.

Interlude: A Winter Night

As a child I was not a bookworm or an infant prodigy: in fact I was rather a tomboy. I climbed trees, collected butterflies, hunted for leeches in dirty ditches, slid down banisters, played practical jokes, fought, romped, inked my clothes. At school, I was rather bright at languages and history, but abysmally stupid at arithmetic; nor was I ever very painstaking at anything. This went on until the age of fifteen, when I experienced a kind of 'conversion', in which (to put it in the most general way) I was given the freedom of a new world of spiritual and intellectual values. I began to read poetry and to speculate about the nature of things. I wanted to go to the university, and with this end in view began to work hard at school. My ambitions were encouraged at home, and being an only child with but few friends of my own age, there was not much to distract me from work. I was intoxicated with a new sense of power, and like a butterfly freshly emerged from the chrysalis, despised the homely natural things from which all this glory had originally sprung. I cast them aside with impatience, as a pettish lady of fashion throws away her last year's hats. I began to learn Greek, not because I was specially interested in it, but because it was something superior, a boy's subject. For even in my most feminist moments I always preferred the interests and occupations of men: they must be more worth doing, or these higher beings would not have earmarked them for themselves. Even now I have a fancy for scything and stack-building, because they are not women's work. I remember an old sailor in one of Mrs Ewing's books, who used to jeer at the girlish business of 'parley-voo and the pianner'. Accordingly I gave up French and, to my intense regret later on, refused to learn music. Possibly Pascal and Beethoven are virile enough to be studied even by the most aspiring of Amazons, but I did not think so then.

In due course I went to Oxford, where I did rather less well than was expected. Two years later, profiting by the delays of demobilisation, I secured a man's job as a university lecturer. I worked very hard, and on the whole was fairly successful; but my ambitious zeal was tempered by fits of staleness and depression.

Experience soon forced me to admit that the equality of the sexes in brain work is a delusion. I found that my male colleagues could do with one hand the work for which I needed two. They had time and energy for domestic life, sport, society: I must concentrate with a single-mindedness only possible for a celibate, and ultimately perhaps only for a monomaniac. There are exceptions, I know; but I believe that my experience is that of very many 'intellectual' women, especially of those over twenty-seven or so. Seeking for relief from the monotony of teaching bored students to turn bad English into worse Latin, I fell into the slough of research, which lies waiting to engulf the brighter and more promising of our graduates. Research, at least in non-scientific subjects, means hunting through innumerable dusty tomes to discover, collect, and present to the learned enquirer a number of facts or theories which are far better forgotten. It is perhaps the most futile occupation in which an able-bodied young person can engage, except perhaps the rolling out of countless bales of cloth for the inspection of women who do not mean to buy a single yard of them.

The intellectual candle began to gutter and smoke; things were working up to a crisis. Research, except the more useful scientific kind that is done in laboratories, is usually conducted in public libraries. The Radcliffe Camera at Oxford is a handsome seventeenth-century dome, fitted up inside with blocks of reading desks, each desk being divided into six numbered seats, which radiate from the centre like the spokes of a wheel. The walls are lined with locked book cupboards; in the very middle of the room, forming a smaller sphere within the circle of the desks, are cases of catalogues, also radiating from the centre. Above is a gallery, full of desks which follow the spoke-like pattern of those below. Indeed, so spherical is the whole building that it gives the impression of whirling endlessly through space at such a speed that its revolutions are no longer perceived. At the desk, students of all ages and both sexes sit side by side, ranged in order like stalled cattle. Outwardly they are mere occupants of numbered seats, intent on gaining information from books. Inwardly, their thoughts and aspirations may be and often are quite unconnected with learning. I once knew an old gentleman with £2,000 a year, who read his Bible every evening at the Camera to save light and fuel at home. I myself have sat at Seat No. K165, trying to decide whether I should or should not marry my man. The number of inward dramas enacted there must be greater than the number of books on the shelves. But these rows of bowed heads appear less personal than the stalled cattle they resemble, because they never betray themselves. The rustle of a sheaf at feeding-time will stir the whole byre: the banging of a much-ordered book upon the desk provokes no movement of delight. Silence reigns, broken only by the hollow boom of closing catalogue volumes, the scurrying of boy attendants with their loads of books, the jingling of the senior assistant's keys as he opens the locked cupboards, a shuffle, a cough, the blowing of a nose. If I have described this solemn place in such detail, it is because I heard in it a more seductive music than the boom of closing catalogues. Pan is no respecter of persons; he plays his impudent pipes in the most august places, in churches, lecture halls, or libraries, wherever there are young ears to hear them. A gale was roaring round the dome, strengthening that strange illusion of whirling speed. Outside, the trees in Radcliffe Square were thrashing in the blast; wild clouds poured over the top of the university church, and smote the western windows of the Camera with sharp volleys of hail. I heard the Pan pipes. Something flamed up in me; I ran down the winding iron staircase from the gallery, across the echoing floor and out at the door, past the friendly old porter with the green baize apron, who used to smile indulgently at the young women students, as if he were wondering what strange desire brought them to this domed temple of dead literature. What indeed! This was the beginning of the end. I often came back to the Camera; but my thoughts not seldom strayed to things which never get bound between the covers of a book.

I had spent my savings on renting and furnishing a vacation

cottage on the Cornish cliffs, where in close contact with Nature and simple daily work I led a life compared with which the academic round of term-time seemed an unreal and rather futile dream. Then came one of those rare and providential opportunities which make even sceptics ask themselves if there is not after all a divinity that shapes our ends. The need for money became less acute: I resigned my lectureship and went to live in Scotland. I was just thirty, strong and healthy, with a good cargo of rubbish to throw overboard, and a clear course ahead. Ten years of varied experience, all of it interesting, some of it painful, has taught me much I should have known long before. But ten years, even in our short span of life, is not a very long time.

I do not think that I have been too severe on these youthful futilities. If I have, it is because everyone who has reached the so-called age of discretion enjoys castigating his own past extravagances; it serves as a measure of progress. The larger the crop of wild oats, the more creditable is the conversion of your waste acres to use and profit. And if my wild oats were not of the orthodox variety, they were wild enough, when judged by the standard of unbalanced extravagance: my pleasure, however sad and un-bacchanalian, was pleasure still, in the carping moralist's sense of the word, which means our inborn love of fiddling while Rome burns. A convert, you will say, always exaggerates his former sinfulness in order to magnify the miracle of his salvation. But to do this you must be quite sure that you are saved. I am most doubtful about it. To the non-Calvinist the boundary between election and reprobation is very hard to find, and sheep almost indistinguishable from goats. No one is assured of salvation, at any rate not till the last hurdle is opened, and the Shepherd with rough firm hands pushes us, bleating and struggling, into the ultimate fold.

It is fashionable to decry culture. One of the most cultured writers of our time has told us why. When everyone is educated, it is distinguished to be ignorant. The summit of distinction would be reached by a Prince of Wales who could not read. But even the most hardened social climber would find something silly and useless in such a disability. We must be educated and we had better be well educated. There are two schoolmasters, life and literature. Life is best, but its methods are very slow, and no leaving certificate is issued till death. Literature is life at second-hand, but for the young it is the only means of anticipating an experience which comes too slowly for early instruction. We cannot do without books, but we can and must do without bookishness. The specialised education that is necessary to equip young people for the professions is bound to be one-sided, especially for girls, whose professional life is often short and always of secondary importance. Bookish education, if it is to be anything but a handicap in life, must be related at every point to practical things. I would like to send every boy and girl of eighteen or so to a new kind of finishing school – a desert island, well wooded and watered, with a soil capable of producing crops and a few edible animals, birds, and fishes. Leave them there for a year with a capable instructor and a few tools, and they would return with a deeper knowledge of human life than could be gained in a lifelong pilgrimage round all the universities of Europe.

When anyone decides to live permanently in the country, they are greeted with a chorus of: 'Don't you find it very lonely?' Personally I do not find it so. There are certainly very few people about, but a crowd is no company, and to the friendless man a big city can be a veritable Sahara. Wherever a few families are gathered together, in the smallest crofting township, there is an epitome of life, containing the whole range of human emotions. Against the changeless background of hills and sea, the figures of men and women stand out in grand, simple lines; their transitory lives and fleeting passions gain dignity and permanence from the eternal natural order to which they belong. One cannot write such words without thinking of Thomas Hardy, who penetrated to the heart of the country, not only in Dorset, but anywhere in the world where men cultivate the soil. But Hardy was a native; he had access to things from which the immigrant countryman is for ever excluded. After many years of patient service the hills and sea may yield the stranger their secrets; the people never. He walks through a Highland village; curious eyes peer at him from veiled windows and through doors ajar; every movement is marked, but he sees no one. A barrier of language, as well as of race and custom, secludes him. With closed doors the natives sit by the fire at their *ceilidh* in Gaelic. The *ceilidh* is a symbol of a kind of fraternity and freemasonry from which the stranger, to his loss as a human being, is definitely excluded. What matter if it is mostly gossip and scandal? Seen in the right light, gossip is the salt of life. Really good gossip can be a work of genius, and it may be more instructive to overhear two old wives discussing their neighbours over a cup of tea than to listen to two statesmen reviewing the world situation over a glass of port. There is something inhuman about the person who gathers no gossip, while he who despises it is either a highbrow or a hypocrite. As for scandal, there is indeed little to be said for it, except that it breeds in its victims a fine carelessness, a dashing disregard for public opinion, which is a healthy and useful quality. Whatever you say or do will be misinterpreted, so why bother? As well be hanged for a sheep as a lamb, and as well for a lamb as for nothing. What say they? Let them say.

5

March: Burning and Ploughing

IN this district, between 65 and 70 inches of rain fall in the year; the monthly distribution varies greatly, but the annual rainfall remains fairly constant over a number of years. There is, however, a wet season and a dry one, though the limits of these are not strictly defined. The drier part of year is from the middle of February to the middle of July, while after St Swithin's Day it usually rains not for six weeks but six months. June is nearly always the driest month, and the wettest may be January, August, October, December, or indeed any month but those of spring and early summer.

In February, when the heaviest rainfall of the autumn and winter is usually over, there takes place a gradual draining of the ground, until the action of the dry March winds, combined with the natural desiccation and withering of all green and succulent herbage, makes it possible to set fire to large tracts of heather and bracken on the hill, and even to burn successfully the rushes and coarse bent-grass in the enclosed pastures.

According to the calendar, March is a spring month; but here you can pass the equinox and feel that the lengthening days do but add an extra load to winter's burden. Gone is the early dusk, when at half past three in the afternoon we can snug things down for the night and retire to books at the fireside. Now, alas! it is possible to work outside till nearly seven, in a stark and lingering twilight without grace or beauty, that lights without warmth, forcing us to work because it is too cold to idle, and because somewhere under the bleached and powdery soil, Nature stirs in her sleep and calls for labour. Birds may be singing and young beasts making growth, but at no time of year does the land yield so little food. On the first day of March, wise farmers will have at least half their keep untouched in the barn.

And at no time of year is there less colour on the hill and in the fields. Frost, snow, and drenching rain have bleached the grass, rotted the leaves, and turned the tawny bracken fronds to powder. If you go to some enclosed and elevated corrie, where the nearer slopes obscure the view of any guiding landmark, with nothing before your eyes but leagues of silent and dun-coloured wilderness where the grass is withered and the burns silent with drought, then you might imagine yourself in any solitude in the world – the African veldt or the Arctic tundra, it matters not where, because the distinctive greenness of the West is to be seen no more. There is no sound but the rustling of dead dry things in a biting pitiless wind, which drives any living creature, man or beast, to the shelter of the nearest rock.

There was much burning to do at Achnabo, and we were waiting with impatience to begin. The right moment is not easily chosen: it may not have rained for a long time, and everything may seem as dry as tinder; and yet the fires will not spread nor burn clean. The air must be really dry, with no morning mist or hoar frost to damp the vegetation. There must be a good breeze, but not a wind so strong that it becomes impossible to control the burning.

We got in a good supply of paraffin, and made a torch out of a short length of electrical pipe fitted with a wick of sacking wrapped round a wire, so that it could (ideally) be pushed up as fast as the flame consumed it. The other end of the pipe was left open for refilling and working the wire. This torch was a doubtful blessing. Either the paraffin came out too quickly, so that a fierce flame ran backwards up the pipe, and the thing wanted refilling every two or three minutes; or it came out too slowly, and in this case the wick got charred and the torch went out in clouds of evil-smelling smoke. It was then necessary to push up the wick by means of the wire inside it. This was all very well, but either the wire was too hot to touch, or it jammed, or worst of all, came adrift from the wick, so that we had to fall back on the tedious and wasteful method of starting a fire with matches and spreading it with handfuls of dry grass. When the torch chose to function properly it was excellent. The burner, with a companion to carry the spare paraffin, would walk rapidly to and fro across the area to be burned, trailing the torch a few inches above the ground; and at his heels there sprang up a line of hissing, crackling flames. As I have said elsewhere, we had no hill ground, and consequently there was no heather to burn; but the pasture land was infested with rushes and bracken, which, together with tussocks of coarse grass left ungrazed by cattle, made excellent fuel for the flames. In these fields there was little danger of the fire extending too far, if only it was kept out of the woods, where tufts of heather, dried needles, and small fallen twigs would feed and spread a blaze that might damage the trees. The trouble was rather that we might start burning too soon, so that the fire would be patchy and spoil any subsequent attempt to burn the field clean. A strong wind springing up suddenly might sweep the flames from end to end in a few minutes; such rapid burning, which only touches the surface, does little good. The best fires are those which spread slowly against the wind; in such burning, tussocks of grass and rush will be entirely consumed, even when standing in water.

Different kinds of vegetation burn in very different ways. When heather is fired, huge flames shoot up with much smoke, roaring and crackling diabolically. Bracken burns with a clear fierce heat; in the moment of destruction, the whole tracery of the fronds is outlined in an incandescent glory, and crumbles into glowing dust. Rushes go up suddenly with a hissing roar like a rocket, and a good deal of sputtering as well, for there is always some dampness lurking at the heart of them. Grass burns with tiny licking tongues of flame, which crawl outward in an ever-widening ring, leaving a charred circle on the bleached pasture. Whins burn with fierce unbridled fury, roaring and shrieking to heaven. No flames so stir and stimulate the wild destructive spirit in man as those that rise from a thicket of these dry and prickly bushes, but the gaunt charred skeletons they leave behind, defacing the landscape for many months, should restrain us from setting fire to them.

This burning of the land is one of the choicest jobs of the year. Everyone delights in making a fire, even when it involves the fierce and exhausting labour of fighting the flames when they have got out of hand. Here we can satisfy our natural love of destroying things easily, almost wantonly, combined with the smug satisfaction of doing good: can be outlaw and Philistine at the same time. For burning not only keeps down harmful weeds and superfluous roughage, but stimulates the growth of young grass and green shoots of heather, bringing an early bite for cattle in the fields and for lambing ewes on the hill. A burnt pasture is about a fortnight earlier than an unburnt one; and tracts on the hill, which for many years have missed the fire, are covered with bushy growths of old heather, which choke the young shoots, and yield but little feeding for sheep in the spring.

Duncan had been out every day with his torch, and our neighbour's hill ground was ablaze in half a dozen places. Shepherds on other farms had also been busy, so that the air reeked with the acrid smell of burning heather. By day, the smoke rolled up in heavy clouds from every hill, veiling the sun and making the higher peaks invisible. By night the horizon glowed with distant conflagrations, while on the nearer slopes were tracts of moving flame, rising and falling in strange contortions, against which the forms of men and dogs were outlined, like the shapes of devils in hell.

Columns of smoke, pierced now and then by darting tongues of flame, poured over the ridges to the eastward, showing that the Laird's shepherd was engaged in his favourite occupation. His hands, white and soft as a woman's, since he never handled a rougher tool than his stick, were grimed with soot and reeking of paraffin; his face was black with smoke and shiny with sweat, and his eyes, habitually mild as a sheep's, glared with a wolfish lust for destruction. He rushed from one slope to another, leaping from tussock to tussock, setting fire to everything without discrimination. He had been ordered to take with him some damp sacks with which to beat out the flames if the fire showed signs of spreading too far. But he had no mind to burden himself with unnecessary luggage; besides, why spoil the fire by trying to put it out? And so, when a wind sprang up and fanned the blaze into an alarming activity, he was forced to use his jacket, and needless to say, it was charred to pieces. The Laird gave him another, which went the same way next year.

A little rain had fallen, and the burning had dwindled into a succession of intermittent bonfires lighted by schoolchildren. The stationmaster sent word that the brooder would arrive by the evening train, and as we had chicks ready to put into it, were in a hurry to get it home. Peter took Dick and the cart to the station, loaded the brooder in its crate, and started on the return journey. About half a mile from the station the road dipped sharply into a valley. The flat bottom of this valley was very boggy, and the road was raised on a bank, which ended in a stone bridge over the stream: on the farther side of the bridge rose a long and steep ascent. The brooder occupied all the space in the cart, and Peter was walking beside it, driving the horse with long reins. They had just begun the descent into the valley, when there appeared on the crest of the hill opposite a line of dancing flames, which in the gathering darkness must have looked to Dick like the end of the world. One glance was enough: he bolted. The heavily laden cart rattled and swayed down the hill, with Peter racing beside it, tugging at the reins. The pace increased; the reins broke, and the boy, struck by the front of the cart, was flung to the ground. When he picked himself up, he found the cart overturned, the brooder pitched on its end several yards away, and the horse lying in a tangle of harness. By some miracle the cart had upset on the very edge of the bank; neither it nor the brooder nor horse nor harness was damaged. It took Peter about an hour to put everything right, and by another stroke of good luck a man came by with a lorry, and helped him to reload the brooder. It was only afterwards, as he led a sobered

Dick past the dying embers of the fire, that he discovered that he was hurt. Meanwhile, as I sat at home with Mrs Gordon, who was still waiting for the cottage to be finished, I began to wonder what was keeping Peter so long; and as the night was fine, I thought I would walk along the station road to meet him. It was very still; there was no sign of burning anywhere, and the actual cause of the accident never occurred to me. I had not gone very far before I heard the rattle of wheels. 'I am sorry to be so late,' he said, 'but we have had a bit of an accident. However, nothing is broken and the horse is all right.' I asked if he was hurt, and he admitted that he was a little. Not wishing to alarm Mrs Gordon, I smuggled him into the kitchen and investigated. He had a nasty abrasion on the thigh and rather a bad shaking.

This was not the only time that Dick bolted. Except in the ploughing, he suffered from excess of idleness, and as an old writer on agriculture puts it, there is no more useless species of bestial than an idle horse. On most occasions he had no objection to the train, but ever so often he would work himself into a panic about a jet of steam, and start off on a wild chase which would nearly wreck the trap on the parapet of the railway bridge. To the hardened motorist, these perils of the road will seem rather silly; but they must remember that a horse, although he may be able to take you home when drunk and incapable, is a living creature, and his behaviour therefore incalculable.

The only real accident I have had with the trap was not the fault of the horse, and may perhaps have been a judgment on me for taking him out on the Sabbath. I was driving alone to have lunch at Strathascaig. In some places the road was carved out of the side of a rock, closely overhung by trees. When within a mile or so of my destination I found that a birch tree had been uprooted by a gale, and was hanging head downwards from the top of the cutting, in such a way that the upper branches were spread half across the road, leaving a space just wide enough to allow the trap to pass. I drove on. The tree must have been hanging by a thread, for in passing my mudguard brushed a twig, and this slight tremor dislodged the whole thing. It crashed down: a branch struck the trap and capsized it, shooting me into the ditch, where I narrowly escaped dashing my head against the wall of rock. The horse dragged the overturned trap about twenty yards, and then with great presence of mind stopped. The shafts being now at right angles to their proper position, it was very difficult to get the horse unyoked; but I got it done at last, and found that the harness was intact and the trap only slightly damaged. The trouble was that it was too heavy for me to lift up alone, so that I was obliged to tie up the horse and go on to Strathascaig for help. This was very annoying, as otherwise I might have hushed up the whole affair. The shepherd and his son were sent to recover the trap, while I went later with a friend to clear the tree, which was now blocking the whole road.

But Dick's idle days were now over. As I have said before, much of the land was in very poor condition. The proprietor, alarmed by the rapid increase of rushes, had ploughed up large acres of old-established turf, and resown it without sufficient cultivation and without a nurse crop. The new grass failed, and the land reverted to bent, sorrel, and other weeds. The spread of rushes, being caused by choked drains and the presence of a hard pan some distance below the surface, could only have been checked by systematic repair and cleaning of drains and deep subsoil ploughing, combined with heavy dressings of lime and a thorough cultivation of the soil. Failing this, the land would have been much better left as it was. The only effect of shallow ploughing was to stimulate the growth of weeds and rushes while destroying the old turf, which was valuable in itself and could have been much improved by the use of lime and harrows. But now the only way of treating this tumbled-down arable was to plough it again and resow it, this time with oats as a nurse crop. Ideally it should have been put through a rotation, begin-ning with potatoes or turnips, but in the absence of labour this was out of the question. I chose two fields: one running uphill from the steading to the belt of wood that separated us from the

Strathascaig hill ground, a long narrow strip of about four acres, incredibly rough and dirty; the other a well-fenced field of five and a half acres with a southern exposure, sloping from the road in front of Rattray's house to the bed of the river. We had also to plough the half-acre strip on which potatoes had been grown the year before, known as 'the black ground'. Thinking that we should be much delayed by adverse weather conditions and the procrastination of the natives, I was anxious to begin in February, but was dissuaded by the Laird, and the first furrow was not opened till 12 March.

I had a good Sellar general purpose plough, but the difficulty was to find a suitable ploughman, one who could bring his own horse and get the job finished before they began the spring work on the crofts, after which time no one would be available. The Laird recommended old Jimmy Macgregor, and after an endless discourse on the goodness of the Laird, the excellence of his own pony, the early hour at which he himself began work, the full value he always gave for a day's pay, the uselessness of my horse, the unsuitability of my plough, the roughness of my land, the hopeless job made of it by all previous ploughmen, he agreed to do the work for 12s 6d a day, he to provide his own dinner and I his pony's. He assured me that he was doing it cheaper for me than he would for anyone else, which I venture to doubt, as 12s 6d a day was the standard rate for a man with one horse. This Jimmy, though he worked a croft and did casual ploughing jobs for the neighbours, had little experience of farm work, most of his life having been spent as foreman of a road gang. 'For forty years', he once told me, 'I never took off my coat!' Which meant, I suppose, that he had directed others without working himself. 'I was never a ploughman,' he said on another occasion, when the horses were taking a spell of rest on a steep slope. 'I would sooner make a tunnel through yon hill than plough this field – it's nothing but a brander of stones!' He had done some contracting on his own account, and might, as he himself admitted, have died a rich man, were it not for the drink. But he chose to live merrily, and was now ending his days on a small croft, the proceeds of which were eked out by various kinds of casual labour. He was a widower, and lived with the four youngest of his twelve children. His house was kept by a daughter of fifteen, and had, as he put it, gone to the devil since his wife died. Peter, who had been inside, agreed that it certainly had; and the Laird, who had known Jimmy for twenty-five years, said that it was there long before that.

He came over on the afternoon of 10 March, which was a Saturday, to look at the plough. It was a good implement, one of the best on the market for our kind of work; but Jimmy shook his head over it. There was something wrong, he said, about the coulter, the feather, the handles; in fact the whole thing was wrong, and I should do well to buy a plough that John the shoe-maker had for sale. I replied that I had no intention of buying another plough, and he must make that one do, coulter, feather and all. He went off in a torrent of eloquence, promising however to come sharp at eight on Monday morning with his pony, and make a start. When he had gone, Peter, who was very handy with any kind of tool or implement, put the plough into good working order, and we hoped for the best.

Monday dawned fine and dry, and as I came out of the byre with a pail of milk for the separator, I saw Jimmy ride into the steading on Polly, a sleek, fawn-coloured Highland pony mare with a tail which swept the ground and a black stripe down the middle of her back. Her owner, a thin, wiry little man with a straggling mous-tache and bright dark eyes, was arrayed in dirty grey flannel bags, bound round the ankle with strips of sacking, broken-down city shoes with very pointed toes, and a ragged knitted cardigan. Hitching Polly to the sawmill rail, he went off to the stable to fetch Dick, while Peter stayed to give him a hand to lift the plough across the road into the field. The horses were inclined to fight, and it took some time to yoke them. At last, when all was ready, Peter and I went in to breakfast. From the house we could hear Jimmy cursing at the horses, and knew that he was under way.

The events of that day, and of several others to follow, brought us to the verge of despair. Every few yards Jimmy would stop, seize a hammer and spanner, and start tinkering at the plough. It had already been set to suit the ground, but this was not good enough; everything must be readjusted, and as Jimmy's idea of undoing a nut was first to turn it the wrong way and then, when it would not move, to bang it with a hammer, the result was not very helpful. Peter went out to assist, and succeeded in preventing him from smashing the whole thing. But progress was lamentably slow. All day the chink of the hammer was heard, and the intervals of cursing, which meant that the plough was in motion, were very short. By way of improvement he took off the furrow wheel, which is still lying beside the fence! By the end of the week he had worn the coulter thin and broken the feather. I had to send for another, which meant the loss of three valuable days, a serious matter in a climate as fickle as ours.

I must admit that poor Jimmy had a good deal to contend with. The field was terribly rough, and on a steep slope, so that the plough had to go up empty. My predecessor seemed to have neither harrowed nor rolled, and the soil was packed into hard corrugations, and plentifully strewn with stones, large and small. Some of these were big enough to have wrecked the plough or thrown it out of the furrows and back upon the ploughman, thus killing him completely (as Jimmy said). The horses pulled badly together; Dick let Polly do all the work and refused to walk straight, so that Peter was forced to lead him. Neither would he turn neatly at the headlands. Jimmy was almost in tears. He assured me that he was ashamed of his handiwork: that if he had had different land, a better horse, another make of plough, more favourable weather, longer reins, and someone to lead the horse all the time, he would have made a far better job of it, and ploughed an acre a day. At the end of the first week so little was done that I began to think we should not have finished our nine and a half acres by midsummer.

However, things were soon on the mend. Jimmy at last began

to realise that the plough worked better when left to itself, and as time went on the cursing intervals got longer and the chinking ones shorter, until he was able to come straight down one furrow without stopping once for repairs! The worst part of the field was done, and the horses were pulling better: Dick had reconciled himself to work, perhaps impressed by the vast range and violence of Jimmy's profanity. This, according to Peter, was terrific, and the sound of it echoed all over the farm. The actual words I never heard, for as soon as he saw me coming he would check the flow, and say in his best Sunday-School voice: 'Come on, Dick! Now, Polly!' One day I ambushed myself in the adjoining wood. As the plough passed, Jimmy was fulminating as usual, but either he was swearing in Gaelic or in English words so lurid as to be unintelligible, for I understood nothing. That day was radiantly fine, and it so happened that a canon of the Episcopalian Church was standing on a hillock nearby, looking at the view. There can have been nothing to distract the holy man's thoughts from heaven but the sight of Jimmy grinding his way up and down the field below, vexing the air with oaths that smacked of some place very different.

The field below Rattray's went much faster; the soil was in better condition and plough and horses working well. Here Jimmy at last allowed Peter to try the plough. He had a poor opinion of boys, and having reared eight sons, he should have known something about them; but he said that Peter was an exception. This may have been his real view, but I have some doubt about it, for he told Peter that he and I could do as much as seven men! Another time, during a short spell in the field, we were discussing Dick. 'A strong horse,' said Jimmy; 'he'll do you twenty years yet.' 'Yes,' I replied, 'I expect he will see me out.' 'Ay, he'll do that, if you work yourself to death as you are doing now. You do far more work than anyone here. The women about here are not much use.' I asked him if that was why he had not married again. He grinned. The eagerness with which this rough, foul-mouthed fellow would rush to save me the least exertion was very amusing. I did, however, persuade him to let me try the plough on the down slope, when everything was running smoothly. I ploughed three rather wobbly furrows, by which time I had had enough. The soil, though smooth in comparison with that on which we started, was rough enough to make the plough rock and plunge like a sailing-boat in a heavy sea. The absence of the furrow wheel made it hard to keep upright, and the length of the handles was a great strain on my arms. Also I had no breath to curse the horses properly. Jimmy walked beside me, trying to cheat by slipping his hand behind mine to support the handle on the left side, which kept dipping down, thus allowing the share to come out of the furrow. Before that, I had tried my hand at every kind of ordinary farm work except ploughing, and having now attempted that as well, I was satisfied.

The rest of the ploughing went without misadventure; the weather remained good throughout, and we had no further breakages; but there was a bad moment when Dick fell down in the plough with all the symptoms of colic. Jimmy kept him walking up and down while Peter persuaded Rattray, who was busy playing the pipes, to go in his car and fetch the Laird. He came at once, gave Dick a dose, and in two days he was back again at the plough.

From what has been said, it is easily seen that the Highlander has not as a rule much aptitude for mechanics. I cannot help sympathising with this weakness. To me, most machines are the work of the devil. What follows here may as well be skipped by the mechanically minded person, as it will only cause him unnecessary annoyance.

Never has the man of science been more in honour with the general public than he is today. The reason is obvious. Most people regard happiness as somehow the gift of machinery, and machinery the gift of science, so that they must owe their happiness to the scientist. There is a flaw in this argument. It is quite possible that machines and their products may make some people happy; but although we could not have them without science, they are not really scientific. The motive of science is the disinterested pursuit of knowledge; the motive of machine-making is laziness, and its methods, like those of arithmetic, could be defined as a kind of low cunning. To the non-mechanical mind, the most characteristic feature of a machine is wheels, and wheels within wheels. The prehistoric genius who invented wheels must certainly have been a very lazy fellow; he may also have been too weak or too sexually unattractive to possess many slaves or wives to carry his burdens. But those who have entered most fully into the fruits of the inspired sluggard's labours are the industrious and thrifty, who use machinery not in order to do less, but to do more. This idea is not acceptable here. Who wants to do more? The more slowly and inefficiently your pot boils, the longer you can spend lounging about and watching it, and thus postpone the awful moment of getting on to something else. Not that the Celt dislikes machines. Practically all Highland fishing boats are fitted with motors, which are used not merely in calms and contrary winds, but altogether, so that the fine art of sailing a small boat will soon be as dead as that of weaving velvet on a handloom. Most young men have motorcycles, for which they are often unable to pay. But few of them show much mechanical sense, and they treat their machines worse than an Italian treats his horse. Unoiled, unpainted, unrepaired, the battered cycles groan and splutter their lives out upon the thankless roads, until they are thrown upon the family rubbish heap, or used to stop a gap in the fence or replace a gate.

All this I can write with a light-hearted feeling of fellowship in misfortune, because, although I have at last learnt to distinguish a screwdriver from a chisel, and once got into the semi-finals of a ladies' nail-hammering tournament, I am one of the unhandiest people imaginable – that is to say, of those who try to do these things. For most clumsy people have the sense to leave tools alone; but I use them, and cut myself and damn and spoil the tool and don't achieve my object. I rarely remember to sharpen anything and still more rarely to lubricate anything. I use a rip-saw to cut across the grain, and do not often drive in a nail without splitting the wood. But it consoles me to think that many people of specialised genius have no manual aptitude. After years of fence – repairing, the Laird's shepherd has never discovered the proper angle at which to lean a wooden stay: neither does he know that in order to fix one piece of wood immovably to another, you must put in *two* nails, or the nailed-on piece will revolve on a pivot like the sneck of a door. And yet he is a marvel with sheep.

I do not own a car, and cannot drive one. Thank heavens! I was going to say, but that sounds rather like sour-grapes hypocrisy. If I could afford a chauffeur-mechanic, I should probably have one. But for all that a car is a diabolical thing. Its brutish obstinacy, its lack of intelligence, its noise, its dirt, its stink revolt the soul. It is far worse to start up a cold engine than to catch a frisky horse. If you are ill or tipsy, you can lay the reins on your horse's neck, and he will probably have the sense to take himself and you straight home. But not even the most expensive car has the homing instinct of a crofter's pony. Relax the wheel, and all it can do is to behave like a sixpenny mechanical toy, and dash blindly on until some obstacle – lamp-post, wall, or bottom of a precipice – forces it to stop. And where are you then? Probably the other side of Jordan, firmly resolved that in the next incarnation you will go back to horses. Of all the humiliations suffered since the Fall by much-enduring man, surely the worst is to lie on your back in the dirt beneath the juggernaut, fumbling in its greasy stinking entrails, while slimy drops of thick black oil ooze down upon your red and sweating face. I once went to the garage at Strathascaig, and found the door locked from the inside. I rattled and called. The Laird's voice, choked with dust and passion and shrill with rage, screamed in reply: 'What the blankety blankety blank do you want? I can't come now.' The scream was accompanied by the dreary chink of some tool. 'He is under the car,' I reflected, and went away, leaving him in torment.

6

March: The Consummation of Winter

BY the middle of March we were beginning to get tired of winter. It became increasingly difficult to keep the cattle in good condition, as our supply of home-grown food was not quite adequate for the number of stock now kept on the farm, and the greatest care and economy were needed in the rationing of hay and straw. Matters were not improved by an unusually late spring: with a few intermissions, winter persisted till within a month of the longest day, while at the spring equinox we were still in the grip of the severest weather of the season. By this I do not mean the lowest temperatures. These occurred during an exceptionally frosty spell in December; but as the days were calm and sunny, the cattle were little affected by the cold and did not require much extra keep. In March, however, the weather, though fairly dry, was harsh and bleak, with penetrating winds. There were broken spells of storm, when icy rain, sleet, snow, and hail came driving on a fierce relentless blast, chilling us to the bone. The worst thing for man and beast is continuous sleety rain, accompanied by a strong easterly or northeasterly wind, and of this we got our share. There was much hail: but a hailstorm is so quickly over and its spectacular beauty so great that the discomfort seems small in comparison with the misery of a blankly wet day.

In winter, each new depression brings with it a warm current from the south Atlantic, which, as the disturbance retreats and the barometer begins to rise, is replaced by colder air from higher latitudes. This change of wind from south-west to north-west is nearly always marked by violent hail squalls, often with thunder and lightning. The swift advance of hail across mountain and loch can be a fearful sight. Below, the storm-cloud is smoky grey; above, a dazzling white or luminous gold, with fringed edges like streaming hair. The whole front is possibly not more than a mile wide. In the middle, the mass of falling stones appears dense as a moving column; everything is blotted out, like a stage when the curtain drops. Sometimes the edge is sharply defined, sometimes blurred and graded, so that the sunlit hills behind are faintly visible, as if seen at an immense distance through a semi-transparent veil. On either side of the squall is transient sunshine and calm water; immediately before it the loch lies black and crisped, waiting for the first gust to churn it into foam. The white column sweeps nearer. From under its moving base, wicked little cat's-paws shoot out towards us. Waves begin to form, but before they have time to develop, their crests are snatched off and hurled headlong in horizontal drifts of spray. Where I stand it is still calm; but I can hear the approaching roar of millions of hailstones striking the water. Nearer and nearer. The inshore islands, gilded for a moment by a vanishing shaft of sunlight, are overwhelmed in white. Nearer still. The middle distance vanishes; the squall rushes across the fields, stripping the surface of the river and flinging it far and wide. The trees lash their branches madly; the rushes are laid flat. An immense roaring fills the air. With a shattering shock the wind strikes the house; hailstones pound the glass like a thousand bullets. Five minutes of turmoil, then calm and sun. A queer world.

One wild night at Strathascaig remains in my memory. A warm day of rain had given place to the usual squally hail showers. The sky was only partially overcast, and the clear spaces were bright with the diffused radiance I have described elsewhere. A squall of hail had just passed over, and there was light enough to see that the head of the glen was lost in a swirling veil of white. Another was advancing from the west, blotting out loch and hills in its passage; but being farther off than its predecessor, and less directly opposite the clear part of the sky, the hail curtain was invisible, and nothing to be seen but vacancy. Spread fan-wise above the main storm–cloud was a trail of mare's–tail cirrus, through which a thin crescent sent down a blurred and sickly gleam. Above this a few stars, clear of cloud but enlarged by unseen vapour, wavered unsteadily, as if reflected in slightly undulating water. A cloudless opening to the south was filled with a continual darting of fierce silent lightning. The squall advanced rapidly, preceded by a calm in which the hiss of hailstones on the loch could be plainly heard. There was a sudden clap of wind, and in a moment everything was drowned in a blinding pelting volley. It all ended as suddenly as it had begun, leaving no sound but the roar of swollen burns.

We now spent a fair proportion of our time in mixing food for the cattle and carrying out hay and sheaves. Three beasts were indoors altogether – the bull, and the two cows we were milking for the house, Aunt Dammit and the Fossil. The bull had hay four times a day, and about six pounds of linseed cake; the milk cows received the same ration of hay, with two feeds of a mixture of barley meal, bruised oats, and bean meal, and one feed of linseed cake. There were eight dry cows, due to calve in the spring or early summer: these came into the byre at night, and had hay or sheaves morning and evening, with an additional feed of hay last thing at night. The five stirks in the heifer shed had hay or sheaves twice a day, with a ration of bruised oats and finely crushed linseed cake at night. The six belted Galloways were out day and night, and received three sheaves each in the afternoon. No roots or chaff were used at all; for maintenance, we depended on home-grown meadow hay which, though difficult to make, was of excellent quality, and on home-grown oats fed in the sheaf. It was not worthwhile to thresh our few acres of corn, and the whole sheaf, combining a concentrated feed of grain with a suitable amount of straw, is, like a Cornish pasty, a useful and easily handled unit of food.

The hay was stored under cover; part of it was in the sawmill, and part in the barn, a long building below the stable. By the end of February the sawmill, so conveniently near the byre, was empty, and we had to carry hay from the barn across the steading and over the bars by the bull-house. We came to the conclusion that it was less trouble to take a small load every day by hand than to refill the sawmill with cart-loads from the barn. We made a carrier, shaped like a stretcher, of light spars of wood covered with sheep-netting, and on this the two of us could carry a large pack of hay. The chief trouble was that our steading, 360 feet above sea-level and completely open to the south and west, was one of the windiest places in the district; and as soon as we started to carry hay, it would blow half a gale, so that unless the load was securely roped most of it would be whirled away into the woods. On washing day the wind always seemed to reach hurricane force; the sheets, though folded in four, would thrash madly to and fro until they came adrift altogether, and, if it were blowing from the south, as was usually the case, they would land in the coal-heap and have to be washed again. Smaller articles would be scattered many yards from the line, while collars and handkerchiefs would often vanish to be seen no more. You will naturally ask, 'Why have the clothes-line so near the coal-heap?' Because the line was stretched between the only two trees in the steading; in the shallow soil lying thinly over rock, no posts could be sunk deep enough to stand the strain. As for the coal, it was lying hopefully at the door of a new coal shed, waiting to be shovelled in when the shed was finished. (Nine months have passed since then; the coal is still there, and the shed unfinished. *Requiescat in pace*.) However, our clothes and household linen are all of the oldest and cheapest description, and get no coddling. Harried by winds, chewed by calves, once even used as rubbing towels by the horse when he broke into the steading, they must take their chance and obey the laws of natural selection. Sometimes in the watches of the night I heard the sheets thundering in the gale, and then after a volley more terrific than anything before, a long ominous silence. Either they had come adrift or the line had parted. Never mind – let them be. I turned over, and thankful to hear no more, fell asleep.

The harvest of 1933 had been secured in good weather, and

the corn was stacked outside. The first stack was opened in December, and about one month's supply was carted to the granary. Stripped of its thatch, the stack had to be re-covered with a tarpaulin, and to secure this against the wind with ropes and stones was a complicated business. One of the worst jobs I know is to open a stack in a howling east wind after a period of frost and snow. The tops of Highland stacks are very steep-pitched and thatched with rushes, which are secured by a network of ropes. Rushes and ropes would be frozen stiff and crusted with snow, and the task of loosening the ropes and stripping off the rushes, perched as you are upon a steep and slippery slope, is no sinecure, and gives one some vague notion of the delights of working a sailing-ship in high latitudes.

The opening of stacks would remind me of Charlie, the youngest son of the Laird's shepherd. Since the departure of Morag, there had been no woman at the shepherd's, and he depended upon the ministrations of this raffish Benjamin and on those of the Slug (of whom more later), who occasionally rolled up from the shore to do his baking. In the long intervals between one household task and another, Charlie strolled on the hill with a cigarette, stalked the hens with his dogs, scoured the roads on a brakeless and lampless cycle, and propelled his father's cow by twisting her tail with the gesture of one who cranks up an old Ford car. In childhood he was lovely as a seraph, and at sixteen he still had the pale ethereal beauty and false fair smile of a fallen angel. Whenever he met you he smiled; whatever you said he agreed with; whatever you told him to do, he replied 'Yes' in crooning dulcet tones, and then did the opposite. Suspected of every kind of devilment, he could never be caught red-handed. So curious and incalculable were his actions, that you were left wondering whether he was a consummately clever villain or mentally defective. When cleaning the byre, he was careful to leave the barrow where it would be tumbled over, and the shovel and brush where they would block a cow's approach to her stall. At one time he used to annoy the maids by shadowing them on

their evenings out: but someone luckily saw a ghost, and supernatural terrors kept Charlie indoors after dark.

Taking sheaves to the Galloways was the best job of the day.
My first experience of the breed was at Strathascaig, and there is
no kind of cow that is more attractive to me. They have a charm
of their own: their soft black coats, often adorned with waves that
would be the envy and despair of a Bond Street hairdresser; their
mild bright eyes, sweeping tails, sturdy legs set well apart, and
most attractive of all, the long feathery hairs that fringe their ears.
They are the hardiest cattle in Britain. Except in unusually severe
weather, when early calving cows are sometimes taken into the
byre at night, they are never inside. A little hay of silage, supplemented with bruised oats and linseed oil after the New Year, is
their winter ration, and they are fed outside. One winter it was
my business to feed them; and every afternoon I would set out
with a bag of oats on my shoulder, calling them as I went. My
appearance was always greeted with a salvo of roars, and the tribe
would come up at a gallop, each to her own box. Late in the
summer, months after all artificial feeding had stopped, I had only
to lift up my voice anywhere in the glen, and a chorus of bellows
was let loose. They calved in February, rather too early for that
cold and snowy season; and for some time the whole family was
taken in at night. The task of sorting out the cows and calves into
their own pens was no easy one: the cows were agitated and
rather wild, while the calves (and few things have less sense of
direction than a young calf) rushed blindly hither and thither,
blundering and tumbling, and in the end getting into the wrong
pen with the wrong cow. The steading would resound with
shouts, bellowings, and whacks; animals and humans slipped and
slithered on the hard-trodden snow; squalls groaned in the trees,
hailstones rattled on the iron roofs. The tumult and the shouting
dies – and it is found that one pen has two cows and no calves,
another no cows and two calves, while two others have wrong
pairs, and only the fifth is correct. So it all has to start over again.
By which time the early dusk has fallen, and the cowherd, with

temper frayed to ribbons, begins to wonder if Galloway husbandry is worthwhile.

Yet when all is said and done, it is the most profitable and least laborious form of cattle-rearing, provided that you have plenty of suitable ground. For the greater part of the year neither cows nor calves need any attention; in the breeding season the bull runs with the cows, and living as they do under natural conditions, the whole fraternity keeps healthy and hardy. Some Galloways are inclined to be wild and difficult to handle, and many are inveterate fighters. One stormy day a farmer found that one of his cows had calved on the open hill. It was blowing a hurricane, and the newly born calf could not get on its feet to suck. The man tried to help, but the mother turned so furiously on him that he ran for the nearest fence. But our Galloways are quiet – quieter than the Guernseys, for they have no private animosities. Even after the grief and excitement of losing their calves, when it was necessary to relieve them of their superfluous milk, they stood motionless to be handled, turning from time to time to look at the milker and lick her sleeve. They move as a herd and live as a family. They are docile and intelligent, wild enough to be charming, and yet sufficiently domesticated to be easily managed. I love them dearly.

It will always be found that cows living under natural conditions and suckling their calves have more intelligence than the specialised dairy breeds. Dairy cows, like hens, whose foolishness is so maddening, are treated so much as machines that their standardised monotonous lives, offering not the smallest scope to resource or enterprise, can be led without intelligence, by the exercise of the lowest form of instinct. We curse them for their stupidity, but it is we who have made them stupid to suit ourselves. The biggest fools, whether among human beings or animals, are those for whom all is found, who need no effort or ingenuity for the getting of their daily food. Thus the half-wild hill Galloway, or the crofter's roughly reared beast, will have more natural brains than the rich man's pedigree cattle. In a neighbouring village where the common pasture is overstocked, many of the cows have learnt the trick of using their horns to raise the latch of the churchyard gate, that they may browse upon the succulent grass within. But the finest example of cows' intelligence is the following. The grazing of a crofting township included a small island separated by a narrow channel in which the tide ran very swiftly. The cows, when put out to grass, used to swim across to this island, and in order to allow for their drift with the current, would enter the water at a higher point when the tide was ebbing, and at a lower one when it was flowing. Thus they had two tracks to the sea, an upper one for the ebb, and a lower one for the flood, and on coming out of the byre in the morning, they would always take the right track, according to the state of the tide!

At Achnabo our Galloways were of the belted variety, known locally as Belties, and had a white band of varying width and shape round their middles. In other respects they resembled the all-black breed, but their calves do not (at any rate locally) make so good a price. They had the run of the fields and adjacent woods, and we fed them in the nearest convenient place where they happened to be assembled. In March we started giving them linseed cake, but they did not care for it, until on Duncan's suggestion we broke it up very small and mixed it with bruised oats. At this time, when Rattray's hoggs were still with us, we saw Duncan every day, and got many valuable hints from him, as well as much pleasant conversation. If you made what he considered a reasonable statement, he would say 'likely' or 'I believe it', whereas any startling remark was greeted with an exclamation of 'Oh be quiet!' Peter and I agreed that we would make Duncan say 'likely' every day until the hoggs went off, and we succeeded, for as he was now buying milk from us, I could always catch him when he brought his can to the dairy. He was deeply interested in the chicks we were rearing in the brooder, and would go every morning to look at them, torn between admiration of the ingenuity and disapproval of the unnaturalness of this method of bringing up the young.

In the intervals of feeding and herding cattle we began to make a garden. I had fenced off a small plot in one of the fields, which had been dug and planted with various vegetables. In spite of gross neglect it had not done so badly, and in the autumn of 1933 there was quite a good show of Brussels sprouts, savoys, and broccoli. But while I was away for a few days' holiday in November, the cows broke in and ate everything to the ground, even uprooting fifty strawberry plants, which would have borne fruit next summer. Farmers are not as a rule good gardeners, and the reason is not far to seek. Garden work must be done out of hours, and is too much like his daily task to have the attraction of a hobby. The man with a sedentary job is much more likely to be interested in gardening, while the farmer in his rare leisure prefers, quite naturally, to frowst indoors. When I earned my living by brain work, I had a passion for open windows; but now, when I am outside all day, my first impulse when coming into the house is to close them, every one! Also, the time when the garden needs the most attention is exactly the busiest season on the farm; and the perpetual vigilance that is required to keep the plot free from marauding cattle, hens, and other livestock whose one object seems to be the raiding of food not intended for them, makes keeping a farmer's garden as it should be kept a very difficult and exacting business.

The garden was first dug by Willie the Scout, an ex-gamekeeper who had served with Lovat's Scouts in the South African War, and now made a living by jobbing gardening and other casual work. His father was a gardener, who made his sons help him after school hours, and taught them the rudiments of his craft. The old man was a martinet, and administered the stick right and left. On one occasion Willie received a thrashing which in strict fairness should have gone to his elder brother. Outraged by this act of injustice, he brooded revenge. He had not long to wait for an opportunity. One afternoon his father put on his best clothes and went to a funeral. The path to the village lay across two fields, which were separated by a deep ditch full of black mud. The ditch was spanned by a plank bridge. When the coast was clear, Willie ran to the bridge with a saw concealed under his coat. Wriggling underneath the plank, he sawed it nearly through, and then retired to the hedge to enjoy his revenge. He watched his father coming along the path, stepping on the plank. . . . All went according to plan, but when he saw the poor man emerge from the ditch, his Sunday clothes all plastered with mud, Willie was smitten with remorse; he sneaked home by another way, and dutifully helped the wrathful parent out of his slimy raiment. No one ever discovered what caused the plank to break. 'I was sorry for him,' Willie concluded, 'but he deserved it. He didn't act fair.'

This field garden was dug again, and in addition we began to make a flower garden in front of the house. This enclosure, surrounded by stone walls 10 feet high, had once been a cattle court, but was now open to the sky and thick with docks, nettles, thistles, and weeds of all descriptions. Under this tangle of matted roots were only a few inches of earth, so we stripped the surface and carted in soil from outside, a long and heavy piece of work, but well worthwhile, as we had a beautiful array of flowers all summer. About the same time we cleared the building called the heifer shed. This had previously been used as an emergency stable, then as a pig shelter, and I believe had not been put in order for forty years. Beside the dried dung, there was an accumulation of rotten wood, old iron, and enormous stones, all of which had to be shifted. It took us three days of navvying with pick and spade to get the place cleared, but we got six cart-loads of wonderful stuff for the new garden.

As Aunt Dammit was newly calved and the Fossil not again in calf, we had plenty of milk, though in order to get it we had to keep both cows indoors night and day and cram them with food. Up till now I had been churning every week and selling the surplus; but I came to the conclusion that it would pay better to keep on Aunt Dammit's calf, and buy in another, thus putting milk into calves instead of into butter, which at 1s 6d a pound does not pay for the making. When selling butter I was interested to find that my chief customers were the crofters; they would walk a mile and more to fetch it, and cheerfully pay the price, while my more prosperous neighbours would not buy it on the ground that New Zealand or Maypole Dairy butter was cheaper. Of course the crofters were accustomed to home-made butter, having their own in summer; and 'shop stuff', however hygieni-cally produced in up-to-date factories, did not appeal to them.

I have spoken of the willingness of crofters to pay a higher price for home-made butter. What do crofters live on? It has often puzzled me. A whole family appears to be able to support life on three or four acres of arable and common grazing for stock, with two or three cows, a dozen hens, and fifteen or twenty sheep, whereas I could not do the same on three times as much land and three times as many animals. It is true that the standard of life, as compared with the south, is low, but there is no real poverty. Never, as far as I can observe, do crofters go really short of either food, fuel, or clothes. How then is it done? I wish I knew, for it would simplify life if I were sure that in the event of a failure in farming I could retire to a croft. I have never tried to investigate a crofting family's budget, and it would be difficult to do so, because so little actual money passes. But I think that the majority of them have some source of income other than the proceeds of their crofts, especially at the present time, when the price of calves and lambs has fallen so low. In most crofting town-ships, the average age is high, which means that a good proportion of the inhabitants draw the old-age pension, thus helping not only themselves but the younger people with whom they live. Many receive remittances from sons and daughters abroad. The younger men will often take temporary work as ghillies, lambing shep-herds, or labourers on the roads, while the girls will sometimes spend the summer in service at a hotel or shooting-lodge. With these aids, and producing as they do the greater part of their food and all their fuel on the croft, they can live in a state of comfort and independence that would excite the envy of many a highly paid worker in town. The crofter is his own master; he can get up when he pleases, arrange his work as he likes, take an outside job if it suits him, and drop it if he has something better to do. He cannot be evicted; if he has a bad season, a benevolent government helps him to tide it over; if he loses his crops, sentimental people in the South collect money to compensate his loss. In handbooks of geography he is described as wringing a wretched living from a stony and inhospitable soil. The soil is certainly stony, but the crofter himself, if only he knew it, is one of the luckiest people alive. What would not most of us give for so much independence and security.

The crofter's daily food, like everything else, is undergoing a change. Potatoes, which were introduced at the end of the eighteenth century, rapidly became the staple food of the Highlands; and they are still the foundation of the chief meal of the day. Milk and eggs are produced at home, oatmeal is still used, though far less porridge is eaten than formerly. Flora never touched porridge, and in our house it was served only to the dogs and to the Sassenachs 'ben' – i.e. in the dining-room. She asserted that it spoilt her stomach, though what spoils the stomachs of most Highland women (and men too, for that matter) is the excessive drinking of stewed tea, black, potent, and syrupy with sugar. When Alec was with us, I used to put two good teaspoonfuls in his little teapot; but when my back was turned, he would add a few more from the caddy on his own account. When Murdo Macgregor came to cut the hay in 1933, he provided his own dinner, which he would take into the kitchen to eat with Alec. Every day he brought two or three ounces of good-quality tea, which he shovelled into the pot; and after the meal his used cup would be half full of thick compound of unmelted sugar and tea-leaves. A cup of tea 'in the hand' is always produced when you call at any house, and the sociable habits of the people must have caused a great increase in the consumption of tea, as also the love of salt herring, which produces the fiercest thirst of any food I know.

There is still much baking of scones and oatcakes on the girdle, which in many of the older houses hangs as in ancient days from a hook and chain over an open peat fire. The newer houses, and those which have been reconditioned, are fitted with stoves, but most of these are semi-open, with a not very efficient oven, so that the girdle is used in preference. A few years before, large hampers of bread came twice a week from Glasgow, penetrating to the remotest corners of the Outer Islands. They still came, but a local bakery had been started, with vans covering the whole district, so that there was no need to bake at home at all, provided that you did not mind eating stale bread. For in remote places the baker's van was seen only once a week, and sometimes not even then, and customers on the country round had to content themselves with Saturday's bread and Thursday's buns. The young driver was far too superior to call for orders at the house; he would blow a blast or two in the steading, and then if no one appeared just turn and go. In any case the time of his arrival was erratic, for one Monday he would take his round clockwise, arriving at Achnabo somewhere about noon, and the next Monday he would go counter-clockwise, reaching the farm between seven and nine at night. He always called at Rattray's first, and as soon as the van was heard shrieking up the brae the Gordons and I went out to the cart-shed, where we sat on logs and awaited its arrival. From the dark depths of the van the baker would rake out loaves with a walking-stick. If we complained of their staleness, he would reply that they were baked that morning, but the wind had dried them! For some reason best known to themselves, the Scotch, who are such excellent bakers of scones and cakes, have no idea how to make bread. The top and bottom crusts of a Scottish loaf are as hard as iron, while its sides and middle are white, soggy, and unappetising. Even Scottish people admit this; and the one thing which the Laird's wife found worthy of praise in England was the baker's bread. I myself am no good customer to the Highland van, for I bake my own bread with yeast which every week comes by post from Hull.

The Highland people are not great eaters of meat, yet they seem able to support a whole army of butchers. Any township has at least one; in the neighbouring village of fifteen or twenty houses there were even two. What actually happens is that a well-to-do crofter keeps a few extra sheep, kills them at intervals, cuts them up, and hawks the meat locally with a horse and trap. Sometimes he buys a bullock from Dingwall and adds a little beef to his stock-in-trade. In addition to these local men, there is even a butcher in Dingwall who thinks it worth while to send a van on a round of nearly a hundred miles, with a ferry to cross each way, in order to supply our small and scattered population with meat. No wonder that the discrepancy between the price of animals for slaughter and that charged for retail meat is so great. With yearling store bullocks down to £5 and beefsteak or gigot up to 1s 6d a pound, we have boycotted the butcher and are living on chicken, macaroni, and ground nuts. I am sorry now that I did not slay my barren cow and salt her down for winter use; we should have lived merrily for a long time, and saved much more than the £1 she made at the sale.

As for the grocer, he never called at all. He thought the road to Achnabo too rough for his van, and I do not blame him. Every Friday he would visit the village below, spending a good two hours in conversation with all the housewives, and Mr Gordon would go down with a basket and get what he wanted. This I was either too busy or too lazy to do, and from time to time I would send to Inverness for a large case of provisions, which came up from the station in the cart.

The opening of this case was a great delight, and the excuse for a feast (or rather for what passed for a feast in our frugal household) which lasted several days. I am always being told how terrible and costly it is not to be able to shop in person. With this I entirely disagree. No doubt the personal shopper, who is able to study the prices of the day, will get bargains which are not accessible to the housewife who orders by post. But how appalling are the temptations to spend money on things she does not really require, seeing that they lie in alluring and provocative array under her very nose, today's bargain, fresh, unique, unrepeatable! She buys something, and when she has got it home wonders what on earth induced her to take it. Whereas the postal shopper, who has to sit in cold blood at her desk and make out an order list, with nothing more inspiring before her eyes than the catalogue of a big store, will think twice before she puts down any fresh item: do we really want it, is it really worth the carriage, will it only be given to the dogs, and so on and so forth, till the list is whittled down to the barest essentials. The extra cost of transport and bargains lost will be more than compensated by the pruning of all those superfluous fallals which we see on the counter and simply cannot resist. The woman who shops by post is freed from the tyranny of the senses, the pride of life, and the perilous delight of going one better than her neighbours. She alone can shop rationally and without psychological disturbance. That is why large drapery firms so often refuse to take post orders in sale time. What value has a sale bargain when you are at home and alone, appraising it in cold blood? None whatever. Thus I have little sympathy for those who complain of the difficulties of catering in the country. It is true that you must think several days ahead, and in terms of hundredweights rather than pounds; but you are spared from the petty tyranny of having to run out every day to buy a bit of this and a drop of that, waiting at counters to solicit the attention of supercilious shop assistants, standing in queues at multiple stores, indicating to some red and corpulent butcher the exact spot at which he is to saw his gory meat in two, and, worst indignity of all, scuttling with chits to haughty cashiers in places where they will not let you touch your goods until they are sure you have paid. What peace descends on the home when the cart has arrived with a sack of sugar, another of flour, and a huge tea-box full of miscellaneous stores! And what a feeling of disgrace, when staying with a friend who was living alone in a flat, to be sent out daily for minute quantities of food, culminating in a demand for two ounces of bacon! May Heaven forgive me for making so foolish a purchase. I nearly went to the pork butcher and ordered a whole pig to be sent round at once.

Meantime the cows' keep was getting shorter and shorter. I had been forced to cut off the morning feed of hay, and buy in a ton of linseed cake, which, according to the Laird, would encourage the cattle to make the most of any grazing they could get outside. The weather remained unseasonably cold, and by the beginning of April it became clear that an extra load of hay would have to be bought from somewhere. All this sounds very improvident, and I daresay it was, but I have something to plead in excuse. Tired out with the harvest, I had taken a fortnight's holiday in November, during which time Alec had run through enough hay to last till the end of the year, and that in a month when little or no extra feeding is required. (No farmer should ever leave his farm, even for a night, unless he wants to find everything wrong when he comes back. This, together with the destruction of my garden by my own cows, rubbed in the lesson.) Cattle prices had fallen so low that I was forced to keep through the winter five stirks that I was expecting to sell out in October as weaned calves. As things turned out, I might as well have sold them, for the extra price they fetched in the following summer barely covered the cost of wintering. And to crown everything, the season was so backward that the young grass came a month later than usual, and we had to keep feeding till the very end of May.

At this time Jimmy Macgregor was coming every day to plough, and, noticing that the barn was getting emptier and emptier, offered me a load of his own hay: it was his practice to make more than he needed for his own beasts, and dispose of it locally. The idea pleased me, as the hay to be obtained from merchants is of inferior quality and high in price, for advantage is taken of the local passion for giving far too much hay before Christmas, thus running things short in the spring. He would not take less than elevenpence a stone, which I thought dear at the time, but changed my mind later. 'Mistress Macleod at the hotel', he added characteristically, 'is paying a shilling, but it will be elevenpence for *you*. And you won't get short measure.'

I certainly did not. Jimmy insisted that either Peter or I should go down to his barn to see the hay weighed. I sent Peter. The hay was forked on to the weighing machine with reckless prodigality, heaped up each time till the scales crashed down without any relation to the weight beyond recording that it was over the measure. The whole load he called thirty-three stones, but it must have been at least half as much again; so that in this deal Jimmy had the satisfaction of making a high price on paper while giving his customer the benefit of something like a bargain! This mass of hay he got somehow packed upon his cart, which was small and had no frame. Jimmy had been down to the village in the morning, and was slightly tipsy, so that the load was built crooked. He called his sixth son, and ordered him to take it to Achnabo. But Murdo, the staidest and canniest of the Macgregors, refused to handle any load so crazily constructed, so that Jimmy, assisted by Peter, had to take it himself.

Peter had left the farm immediately after dinner, and Jimmy's croft was not more than two miles away. By half past six there was still no sign of the cart. The milk cows were grazing on the way to Jimmy's, and I reckoned that if I went out to fetch them home, I should meet the cart on the road. Nor was I mistaken. I had just reached the gate of the cow pasture, when I heard the grinding of wheels over the rough stones, and looking down the road, which at that point began to drop sharply downhill, I beheld a monstrous object labouring up the brae. The toiling pony was almost hidden by that vast and uncouth load, which had developed a heavy list to starboard, and as it brushed the overhanging branches of trees, threatened each moment to fall and bury Jimmy, who staggered beside it, urging the horse with language that was luckily drowned by the rattle of ungreased wheels. Of Peter, who was presumably steadying the load from behind, there was no sign. 'You've got a big load there,' I remarked. 'Ay,' said Jimmy with a grin, 'it will last ye a fortnight whatever.' I made up my mind that it should last a good bit longer than that, and it did; it was good strong hay in first-class condition. The cart had now reached the level, but the road at this point was full of pot-holes, some of them six inches deep, and as the cart lurched and swayed along, my heart was in my mouth, for it was obviously going to rain in a few minutes, and if once the load fell off it would be no easy task to fork it on again. But fortune was kind; we reached the farm without losing a handful, and had thrown most of the hay into the sawmill before the heavens opened and we were drenched with one of the heaviest downpours of the season. Jimmy was as usual impervious to the weather; he scorned a coat and refused a cup of tea, asking only for leave to take away an old wheelbarrow, so useless and decrepit that I had earmarked it for our next bonfire. I gave it gladly, thankful to see the last of it. He hitched it on to the cart, and as he turned away remarked that the man yonder had been after his daughter for a servant, but he had forbidden her to go; he would sooner see her in a tar-barrel than there. It was probably as well for her prospective employer that the youngest Macgregor was not allowed to take service with him, for domesticity was not the strong point of any member of that family. It may have been the discomfort of his own home that made Murdo hanker after something different, for I once heard him and his crony Alec discussing the domestic perfections they would demand from their wives. 'If you want girls like that to marry you,' I said, 'you'll have to work hard and earn a lot of money.' 'Ay, we mean to do that,' they replied. Alec's idea was to run a market garden, while Murdo, who did not care for digging, proposed to save up enough money to buy a stallion, and tour the country with it! So far, neither has realised his ambition.

7

April: Planting and Sowing

No sooner had Jimmy started his first furrow at Achnabo than he began to attack the problem of our potatoes. He would ask me almost every day where I was meaning to plant them, what method I should adopt, how many drills I wanted, how much dung, how many helpers, and so on and so forth, all combined with such an outpouring of advice, information, and reminiscence that I would hastily flee to the house, so as to spare my ears and let him get on with his work. The solemnity and awe which in the Highlands surround anything to do with the cultivation of potatoes is at first rather perplexing. This uninteresting and comparatively modern crop has gathered to itself all the romance and traditional glory which from the dawn of history has attended the sowing and harvesting of corn. On the crofts, the great spring festival, when people meet for mutual help in work and talk, is the planting of the potatoes, while in autumn the whole year's labour is brought to a close with the solemn ceremonial of the lifting. This may be because home-grown oats are no longer used directly for human food; oatmeal is bought from a miller, and the crofter's own corn is fed to his cattle. But potatoes are the staple diet of the people, and have been so since the latter part of the eighteenth century. They appeared in the Outer Islands as early as 1743, when Clanranald, the proprietor of Uist, introduced them from Antrim. Dr Walker, whose book, *The Economical History of the Hebrides* (1812), should be read by anyone who is interested in the traditional agriculture of the Highlands, gives an amusing account of Clanranald's experiment. 'The tenants were convened and directed how to plant them, but they all refused. On this they were all committed to prison. After a little confinement they agreed at last to plant these unknown roots, of which they had a very unfavourable opinion. When they were raised in autumn they were laid down at the Chieftain's gate by some of the tenants, who said that the Laird might indeed order them to plant these foolish roots, but they would not be forced to eat them. In a very little time, however, the inhabitants of South Uist came to know better, when every man of them would have gone to prison rather than not plant potatoes.' And so it has remained, and the crop has gained an importance out of all proportion to its intrinsic value as food.

Potatoes have the advantage of being easily grown almost anywhere in the Highlands, and since the introduction of immune varieties, can be trusted to produce a fair crop in almost any season. They respond well to the manure most easy to obtain locally, dung and seaweed, and are very tolerant of lime deficiency. In the old days, when crofters' land was never ploughed, but dug by hand with the *caschrom* or crooked spade, they were planted in 'lazy-beds', which were made as follows. A grassy spot was selected, and trenches were dug at intervals of a few yards, which served as drains. The space between the trenches was covered with a layer of seaweed, rotted bracken, weeds, cut grass, etc. On this the seed potatoes were placed, and the whole covered in with earth taken from the trenches. Potatoes grown in lazy-beds seem to have yielded considerably less than those planted in drills; but as drilling was practised on gentlemen's farms, where both seed and general management were probably better, it is difficult to come to any definite conclusion. Lazy-bed cultivation was an easy method of bringing wild land into profit, and traces of it can still be seen on many a hillside which, with the decline of the crofting population, has reverted to its original condition. At the present time, when every township has at least one plough team, the usual practice is to dress the land liberally with dung and seaweed, plough it in, and plant the potatoes in the furrows. For this work every person in the place is mustered, men, women, and children. When the plants are well above the ground they are kept clean by repeated hoeing, which is generally done by the women.

Country women in the South would be horrified at the amount of outdoor labour which falls to the share of the wives and daughters of crofters. In the old days, men who had no ponies attached the harrows to their womenfolk, or loaded them with heavy creels of seaweed or peats. 'The burthen of work', says an old writer, 'was cast upon the females. The men deemed such an occupation unworthy of them, continued labour of any sort being most adverse to their habits.' Now that iron harrows of the standard pattern have superseded the antique implement, which was really an enormous rake with wooden teeth, ponies are employed for this work, but there is still much carrying of heavy loads. I have sometimes seen women of 70 and upwards with creels of peats on their backs: in one case the old dame had a grown-up son at home who owned a horse. One hot summer's day in Skye I watched a characteristic scene. The woman was hoeing potatoes, while her husband sat peacefully on a boulder, smoking his pipe and watching her at work. The skilled and pleasant jobs – ploughing, scything, wood-cutting, the tending of sheep – are in the hands of the men, while the dull, heavy routine work is done by women. Hardly a man knows how to milk a cow; the making of hay and binding of sheaves is a woman's business; while poultry-keeping in all its branches is looked down upon as only fit for grandmothers. When the Laird took over the care of his fowls, and was seen walking down the steading in kilt and gum boots, scattering grain and calling 'chook! chook!' in a piercing falsetto voice, followed by tribes of hens of every breed and colour, he must have caused the greatest scandal of the century. I once told Flora's brother that in England the women rarely milked, while the best poultry farms were managed and staffed by men; but I do not think he believed me. Old notions die hard. In ancient times, when men were mostly away hunting, fishing, or fighting, it was natural enough that the women should cultivate the land. But now, when these activities have been so much curtailed that man's chief work is leaning against the door-post of the cowshed, talking and smoking until something turns up to interest him, this unequal division of labour has little to be said for it. Houses are badly kept, cookery hardly exists; no one bothers to keep a garden, much less to make clothes. A certain amount of knitting and spinning is done, but not one of the Highland girls we had as servants brought with her a needle and thread, nor did I ever see any of them darn as much as a stocking. Nor can you blame them; they have too much to do outside. And there may be more serious consequences. Some years ago, after a spell of bad harvest weather, there came a fine day or two, followed by a night of brilliant moonshine. Anxious to get the corn in, a crofter's wife worked all night. Her baby was born prematurely and she herself died, leaving a family of small children.

To go back to the potatoes. I had fixed on a plot at the bottom of the long cornfield, sloping to the road. Needless to say, this decision did not find favour with Jimmy, neither did my plan of planting only eighteen short drills, just enough to supply ourselves and the Gordons. Still less did he like the idea of drilling them at all. However, he promised to do what he could, and on 17 April he came up to see when I intended to make a start. Peter had been busy grubbing and harrowing, in order to get the plot into some semblance of smoothness before the drills were opened. But there had been heavy rain in the night, the ground was very wet and impossible to work fine, so we postponed the planting until a period of drier weather should have given us a chance of making a finer tilth. Actually the potatoes did not go in until the 26th, the day after we finished sowing the big cornfield. From the 17th to the 23rd the weather was very broken with frequent showers of heavy rain, and we did not get much cultivation done, especially as it was difficult to borrow a second horse at this time of year. So that Dick had to work alone, pulling two instead of three harrows, and he made a poor job of it. Not only the potato plot but the whole cornfield was terribly rough, full of large unbroken divots with deep pits underneath big enough to have buried whole pounds of seed. After the 23rd a fine spell set in, during which we got the big cornfield sown and harrowed; but of this, more later.

We had always planted potatoes in drills at Strathascaig, and as I had a drill plough on the farm I decided to do the same at Achnabo, Jimmy or no Jimmy. The procedure was to harrow the ground fine, open the drills, put dung in the bottom, lay the potatoes on the dung, and close the drills by splitting the ridges with the plough. As Jimmy had so rooted an objection to the drill plough and all its work, I asked Hector Macleod, who came on the 25th to sow the corn, if he would also open eighteen drills for the potatoes. This he did that evening.

The 26th dawned fine, with a light northerly wind. About ten o'clock Jimmy, accompanied by Murdo and Polly the pony mare, arrived to plant the potatoes. Jimmy inspected the drills made by Hector and shook his head over them. 'You will never get a crop in the like of that,' he said. Meanwhile Peter and Murdo carted dung from the midden and threw it out by forkfuls into the drills, while Jimmy and I spread it evenly along. The roughness and softness of the ground made the carting very heavy, and only small loads could be taken. Jimmy kept up a continual stream of lamentation; the drills, he said, were too deep, and the potatoes would never come through. With the idea of making them shallower, he started grubbing about with a fork, and succeeded in so far knocking in the sides of the drills that for all practical purposes they might not have been there. 'You would never have got a crop with yon.' By this time the very thought of potatoes had become hateful to me, and I testily replied that I didn't care a hoot whether I got a crop or not, at which rank heresy Jimmy stared in pitying wonder. Finally the bags were carted out, Mr Gordon arrived, and we all began to plant. A heavy shower at dinner-time caused some delay, but the wind soon dried up the surface moisture, so that the planting was finished by 2.45 – three drills of early varieties, and fifteen of Kerr's Pink maincrop.

It now remained to close the drills, or what was left of them. Jimmy said it was impossible, but nevertheless he would try. Murdo, who had been very tactful and philosophic all day, grinned at Peter, and remarked that his father was getting a bit old in the head. They yoked the horses to the drill plough. We thought it best to leave them to wrestle with it alone, so Peter and I retired to the flower garden, and planted pansies and Canterbury bells. The high walls prevented us from seeing the field, but the continual stream of bad language kept us pretty well informed of what was going on. The horses were lurching this way and that, knocking potatoes out of the drills and trampling them underfoot; the plough was splitting something, but whether ridge or furrow it was difficult to say. At this point Jimmy scrapped the drill plough, and tried the single plough instead. This, it seemed, would work no better; and at last he came to me in a state of prostration saying that it was impossible to go on: we must gather up the potatoes, re-plough the plot, and plant them again without drills! To which I replied that we would do no such thing; he must just carry on and get finished somehow. The real trouble was that the plot was beside the road, and he was afraid that passers-by would see the mess we had made and laugh at him. In the middle of the operations the post had come by, and would of course broadcast the dismal failure of the Achnabo potato-planting, and Jimmy would be for ever discredited. I tried to console him by pointing out that I should be the person blamed, as it was I who had ordered drills to be made. At last the plough was set going again, and by six o'clock the majority of the tubers were hidden from view, though where or how they were lying no one could tell. As a result of Jimmy's levelling of the ridges with his fork, the potatoes were so thinly covered with soil that I dared not follow the usual practice of harrowing down the drills three or four weeks after the planting; and when the plants did actually struggle through, the rows were too crooked to be cleaned with the horse-hoe, and had to be done by hand. Apart from the satisfaction of producing things ourselves, it would have been far easier and cheaper to buy potatoes from a neighbour. But even with all the neglect and misfortunes suffered by this crop, it served us well, and justifies my theory that (apart from scientific production) you can grow potatoes almost anywhere and almost anyhow, and that the detailed and semi-religious attention that is given to them here is rather a waste of time. I had degenerated seed, wretched cultivation, no potash manure, no harrowing or horse-hoeing, inadequate hand-hoeing, and on the top of all that a summer and autumn so wet that not a few crops in Skye rotted in the ground before they could be lifted. And yet I have in the barn a store sufficient to take us through the winter. The potato, which forms the staple diet of the Aran Islands and of Tristan da Cunha, cannot be a delicate plant. Why then (unless you are a scientific grower) trouble to coddle it?

The corn was sown in three sections, beginning with the 'black ground' and the lower part of the long field, which were seeded by Jimmy on 19 April. In this way I hoped to avoid having the whole crop ready for cutting at the same time; and although the last sowing nearly overtook the first, it was I think worthwhile. Jimmy turned up unexpectedly early, thereby causing much confusion, as we were still at the milking and had not yet had breakfast. For some unknown reason he had failed to bring the mare, so that Peter had to post off to the village to borrow a horse, while I seized upon Flora, who was dreamily stirring the porridge, and sent her to the field to serve Jimmy with seed. She was presently relieved by Mr Gordon: then Peter returned with an antiquated black pony, which looked far too small and frail to pull with Dick. A rapidly increasing wind made it difficult to broadcast evenly, and by dinner-time all work was stopped by a deluge which lasted for the rest of the day. But all the seed was covered before the rain came. It appeared that Jimmy's mare would no longer be available, as the spring work had started on the crofts. We must therefore get another man to finish the sowing – one who could bring his own horse; and after much discussion at Strathascaig on the following Sunday, it was arranged that Hector Macleod should come to us on Thursday next if the day was good.

The day was very good – bright and sunny, with a light northerly wind. Hector arrived early with a fine horse of Dick's size and type. This young man was the son of a small farmer, who on his father's death had come home to run the farm for his widowed mother. He was quiet and competent; for the first time in the history of our spring work did we get through the day in peace, without hitch or blether. Bags of seed corn were carted out and deposited at convenient intervals; Hector sowed, while Peter followed with the harrows. I was kept busy carrying pails of seed from the bags to the sower, a more strenuous job than it sounds, for a two-gallon pail full of grain is heavy, and the ground was very rough. Also it is not always easy to fill the pail without spilling. Hector was a beautiful sower. Up and down the field he went, with long deliberate strides, his arms moving rhythmically in wide sweeps, scattering the seed right and left, so that it fell evenly like drops of rain. This is a dying art: few people have seen corn sown by hand, and fewer still, when listening to the wireless, stop to consider what broadcasting really means. To broadcast well is not easy, and my desire for a good crop would not allow me the pleasure of trying it myself. At Strathascaig I would sometimes steal the canvas sowing-box, sling it round my neck, fill it with wheat for the hens, and stride solemnly down the steading sowing it for them as I went. Later, at Achnabo, I had a few hundredweights of nitrate of soda to spread on one of the pastures. The crystals were about the same size as grains of corn, and seemed suitable material with which to practise a broadcast. The white grains were easily visible on the short grass, and I flattered myself I was doing it rather well, especially when Duncan, who was passing by, stopped to watch me, and said, 'Ye've an awful notion of the sowing.'

Hector finished the long field by dinner-time, and then proceeded to sow the 'black ground' with grass, for which the corn previously sown by Jimmy was to serve as nurse-crop. The mixture I had chosen consisted half of cocksfoot, and the other half of fairly equal portions of rye-grass, timothy, hard fescue, red clover, and alsike. Luckily the day remained calm, for grass seeds are so light that the least wind will make them fall unevenly. Grain can be sown with the whole hand, but the airy grass seed must be taken between finger and thumb, and scattered with the greatest care. I had intended to sow the long field with grass as well, but even after repeated harrowings the ground remained so rough that a good germination seemed impossible: the idea was therefore abandoned.

The field below Rattray's was not sown until 3 May. This field had been well ploughed and repeatedly harrowed, and was in fine condition. Everything went like clockwork. In the middle of the morning a small urchin came up with the cow for the bull, and Peter had to leave the harrows to attend to her. By six o'clock it was clear that by working late we could finish that night. We kept Hector for tea, and persuaded him to leave his horse overnight, as Peter could finish harrowing in the grass seed alone, provided that the second horse was left with him. Hector agreed to this, and went home himself about eight o'clock, leaving Peter to finish the job alone. Summertime had begun, and the evenings were already very long; and when the horses were finally unyoked at ten, it was still quite light. Thus ended the most satisfactory day's work we ever did on the farm.

The cows were still indoors at night, and on retiring from the field I went into the byre to look at Priscilla, whom we expected to calve any time now; but as she was a sucker and had been running with the bull all the previous summer, we did not know the exact date. From her appearance I thought it not unlikely that she would calve before morning; but as Priscilla had never had any trouble with her calves, and we were both incredibly tired, we left her and went to bed. Next morning a black and white heifer calf was skipping about the byre.

In the intervals of sowing we had put up a temporary fence to keep the cattle off the 'black ground' until after harvest. Murdo and Peter had previously cut some posts in the wood and carted them to the 'black ground', but it was Willie the Scout who helped to make the fence. I had asked him to come over on the first good day; but as the first good day was cloudy in the morning, and Willie's house was so deeply embowered in trees that only a small section of sky could be seen from it, he assumed that it would rain, and went elsewhere. It did not rain; and when he came at noon he could have bitten his fingers off, as he put it, for being so poor a prophet. He looked at Murdo's posts, shook his head over them, and gave Peter a demonstration of post-driving. They also dug in the strainers and stretched one wire. That was on 23 April; in true Highland fashion the remaining wires were not put on till the 29th, and then only because George, spying some of last year's potatoes which had been thrown up by the harrows and left lying at the side, pushed his way under, slackening the whole wire and loosening three or four of the posts. These posts, Willie had assured us, would never move. Now everything that Willie fixed would, on his own showing, last till the Day of Judgment, and yet there are few gates and posts of his that I have not seen shifted by some beast or by the mere process of natural decay; and I have not known Willie for more than six years. He was a splendid worker, but a streak of obstinacy and impatience spoilt everything he did. He dearly loved showing women and young people how to do things, especially things that required some spectacular exhibition of strength. 'Look you', he would say, and, spitting on his hands, swing his fourteen-pound fencing hammer with that overhand stroke that always looks so masterly, and bring it down square on the head of the post. I must have had little faith, for I always hated steadying the post for him: what if he missed the top and hit my fingers? He never did, but still . . . As for Peter, he knew how to do most of the things in which Willie delighted, and many of them better, so that the 'look you's' fell a little flat. What Willie really liked was a feminine audience. I remember seeing him in the midst of an admiring crowd of girl haymakers, showing them the right way to build a coil of hay. 'Look you!' he said, beaming with delight, as he shaped the thing to a delicate point. 'It will never take the water – not if it stays here all winter!' He was careful never to swear in the presence of ladies, and when he cut his finger on a sharp end of wire, the most he would say was 'Dash it!' 'Ach, say Damn and have done with it, Willie,' I said to him once, 'there's no ministers about here.' But for all his little weaknesses he was a dear friend, and I wish I could have afforded to employ him more often.

No sooner was the corn sown than hundreds of seagulls came swooping down and settled on the field, walking over it with their clumsy waddling gait and picking up grains. It all looked very destructive, but actually they did little damage, taking only the grains exposed on the surface, which in any case would never have germinated. Peter went out with his gun and scattered them: one bird was brought down and left in the middle of the field as a warning, after which the whole tribe rose high in the air, wheeled screeching above the field, and vanished to be seen no more.

The shooting of gulls served as a cloak for other activities. When I first came to Achnabo there were two cats in possession, one a lean grey and the other a rather less lean tabby. This pair were assiduous hunters, and were out for days together, only returning to the farm when their outdoor store of food was exhausted. They were fed officially by Flora, but so great was their skill in thieving that nothing in the house was safe. The lean grey one was the worst, and after dining once too often on silverside at eighteenpence a pound, paid the extreme penalty. Charlie, the tabby, remained, and spent the intervals between his hunting and courting expeditions in repose on the kitchen window sill, from which vantage-point he would slip in now and then and raid the larder. We were getting rather tired of Charlie. On 24 April he ate Murdo's dinner, which had been left in his jacket pocket in the cart-shed. On Sunday the 28th, the weather was so beautiful that we took lunch into the woods; on our return we found the leg of mutton lying half consumed on the kitchen floor. This was the last straw. Next day he was executed at dawn and buried in the corn-field: everyone except ourselves believed that Peter had been shooting gulls. We had to burden our conscience with many lies, for the Laird's wife, who was devoted to cats, used to ask us every Sunday how Charlie was. I fancy she suspected something, but we always asserted that he was away somewhere, we did not know where. Nor did we.

In the last week of April the lambing began. This did not concern us very nearly, as we had no sheep; but Peter went out several times with Duncan on his morning round, covering a great stretch of hill in a comparatively short time. The season was unusually backward, and the shepherd was very anxious about the lack of keep for the ewes. A new-born lamb can stand any amount of cold provided that its mother has plenty of milk. But if a protracted winter has delayed the coming of the young grass, the ewes have little milk and the death-roll is heavy. In hill flocks, most of the ewes lamb on the open hill, only a few of the less hardy being brought down to shelter. No feed is given, and the shepherd's chief desire is for warm showers to hurry on the grass and young shoots of heather.

I feel ashamed that this book, which is all about farming in the Highlands, has so little to say about the Highland farmer's chief source of income. But this is not altogether my fault. It is one of these odd disabilities that crop up in the most unexpected places, and go to prove that sex discriminations not based on biological differences are mostly irrational prejudice. In the Highlands, cattle are chiefly tended by women, while sheep are an exclusively male concern. A shepherdess is no more than a literary puppet, and religious superstition has been careful to cumber the woman with skirts to the ankle, thus making it impossible for her to walk on the hill. Why? Probably because herding is on the whole a pleas-ant job. How sweet is the shepherd's sweet lot! You will say that Blake never thought about the messiness of dipping, which in his day was not forced upon us by the government; but that only happens twice a year, and, for the rest, the shepherd has the nearest approach to a gentleman's life that I know of. He has, it is true, his rush seasons, the heavy labour of clipping, the anxiety and weariness of lambing, but in the long intervals between he can take his ease. He need not dirty his hands, nor soil his clothes, nor stoop nor run about; he whistles and shouts and points with his stick, and the devoted and intelligent collies do the rest. He has a house, cow, hens, meal, and potatoes innumerable. What more can he want? He is a gentleman. As for his master the sheep-farmer, he has the nearest approach to unlimited ease of anyone who claims to work at all. He has the minimum of labour to supervise, and the minimum of occasions that must be personally attended to. He must know something about sheep, and be able to buy and sell to advantage; but when he has visited two or three sales a year, and presided over the clipping, dipping, and sorting of sheep and lambs, he can devote the rest of his time to sport or any other personal hobby; if he has a good shepherd, he can even afford to be non-resident. But he must have plenty of capital, a shrewd eye for business, and either genuine experience or the capacity to feign it. Yes, the shepherd and his master are gentlemen; but I am only a woman, and so I know nothing of shears, collies, dip, keel, cast ewes, gimmers, hoggs, earmarks, and all that. I would like to have travelled with that Patagonian pioneer who went so far to seek new pastures for his flock that his ewes lambed twice on the journey. But I have never done more than take up tea to the clippers, and, being a fast runner, act as collie substitute at the fank. So that when sheep appear in these pages it is mostly in a vague literary way, for which I apologise, while protesting that it is not my fault.

For in any case, however much I might break down the traditional barriers of sex, I could not keep sheep at Achnabo, because there is no hill ground. It would have been possible, and probably profitable, to have kept a small number of cross-bred ewes on the enclosed fields, but these must have been bought in from outside, and owing to the general wreck of the fences I had not a single field in which they could have been shut up until they settled.

There were plenty of sheep on the farm, but none of them mine, for the fences which would not keep in my sheep obviously would not keep out anyone else's. When Rattray's hoggs were taken off at the end of the first week in April, their places were immediately taken by crofters' sheep; and at one time we counted twenty-eight ewes and about forty lambs getting a free living off one of the fields we had shut up for hay. I had long since abandoned the more distant pastures to these marauders, as the fences there would not have held up a rheumatic old woman on her daily walk, far less an athletic ewe in search of fresh grass. But I made a heroic attempt to protect the fields nearer the house, more especially those laid up for hay or carrying a crop. We closed every possible gap, and patrolled the place several times a day, clearing out the intruders as best we could without a trained dog. Thos, my mongrel collie-cairn, could put out sheep quite well if he chose, but was generally too lazy to chase them far, and was liable to run for his life if some courageous ewe turned to face him. And after a time they got so hardened that they paid no attention to the dog, and even if they were moved on a little they would be back again in a few minutes.

There is no more active or persistent beast than a crofter's sheep, or his cow, for that matter. After speaking of the careless disregard of property shown by the natives, an old writer on Highland agriculture continues: 'Nor is this strange tendency to hurtful activity confined to what we call (perhaps Hibernically) the rational animals of this country. The horses, sheep, and cows are universally of similar disposition. The same enclosure that suffices for protecting the rich meadows of Suffolk and Essex would no more be heeded by a Hebridean beast than if it consisted of the mist of the mountain.' This I found to be as true at the present day on the mainland as it was in the islands a hundred years ago. By the middle of May the whole place swarmed with sheep, mostly from two adjacent crofting townships, with one or two of Rattray's, and about half a dozen of the Laird's, including a most persistent ewe with twin lambs, who squatted permanently on the young corn in the 'black ground'. Periodically I got the owners of some of the sheep to round up their flock and drive them up to the hill, but within twenty-four hours they were back again. Nothing short of a ring fence of sheep-netting would have kept us permanently protected, and that was utterly beyond the means of a struggling tenant farmer. We were constantly on the run; I even dragged Rattray out of bed in the early morning to send his dog after some sheep, believed to be his, that were eating down the corn. We remonstrated with the Laird, who sent over his shepherd to gather the trespassers, but when he arrived every Strathascaig sheep had miraculously disappeared. 'There will not be much for them here whatever!' he remarked critically, looking at my field. Within an hour of his departure they were all back, so I imagine that they must be finding something there; however I asked the Laird if he thought that half an acre of young oats would keep a ewe and twin lambs for a fortnight. He grinned, and later on suggested that after the Belties had been taken from the bull at the end of July they would be the better of a spell on his hill ground. They went, and stayed there till nearly the end of the year, to the great improvement of their condition, so that we did rather well out of the Laird's trespassing. But there was no compensation for the other marauders, and they spoilt a good half of one of our hayfields. As things turned out, this was a smaller loss than might have been expected, for the weather in September was so abominably wet that we were unable to cut this section at all.

We were greatly handicapped by having no trained collie. Thos was better than nothing, but he was a far from adequate helper in these emergencies, and I began to make enquiries for a young dog that had had a certain amount of training. I heard that the butcher at a village nearly three miles away had a suitable collie, and one afternoon the young man rode up on his pony to see me, bringing the dog with him. It was a nice young beast, and the butcher demonstrated its working capacities on some stirks that happened to be in the sawmill field. We had a prolonged haggling match, in which I reminded him of the many gigots I had bought from him for the haymakers the previous summer, and of how I had recommended his meat to the Gordons, and so on and so forth. 'I know you have been very good to me,' he said with a charming Highland smile, and knocked off ten shillings. I agreed to take the dog on a week's trial, and it was shut up with Flora in the kitchen, where it howled the house down and nearly smashed the window in its endeavours to get out. As he rode off, the butcher's eye fell on a derelict bath at the end of the sawmill field, lying idly on its side, just as it was left when the white bull had finished playing with it the previous summer. He asked me if I were using it, and if not, could he have it to make himself a dipper. I replied that I had no use for it, and he could cart it away with pleasure at any time. A few days later he called and took it away, as well as another which some lunatic had fitted up as a drinking-trough within a few yards of a large freshwater loch! After a night of howling, the dog gave us the slip and went home. We made another attempt to acclimatise him, but it was useless, and we were without a collie until Murdo Macgregor presented Peter with the puppy Donas Beg.

8

Interlude: The Countryman Born and Made

MANY of my friends are Catholics, and I have never ceased to notice the difference between the born Roman and the convert. New brooms sweep clean; and apart from that, to change one's religion is a difficult, often a painful business, demanding much zeal for its accomplishment. In Britain the proportion of converts is high, and it is they who are responsible for the silly idea that all Catholics are engaged in ceaseless propaganda. There is always something a little hectic and restless about any kind of convert. But the born Catholic mostly takes things for granted without fuss or question, and talks but little about his faith. On the surface he appears indifferent; he is at his ease in Sion, and shocks the earnest Protestant by his lack of seriousness.

So also the born countryman. He speaks little and writes less about 'the country' as such. When he walks the fields, he is not in search of sunsets or uplift, but looking to see if his heifers are in good condition or whether the hay is ready for mowing. Natural beauty is part of his daily life, the very air he breathes; he is not consciously aware of it, any more than the healthy man is consciously aware of his digestive processes. Those who talk loudest about beauty are the people who cannot get enough of it, just as celibates have most to say about sex and unemployables about labour. Since I have come to live in beautiful scenery, I feel far less desire for books about it. What need would Henry VIII have felt to read or write about the pleasures of marriage? Long ago, when I visited Cornwall as a tourist, I was delighted by some effect of light and colour on the sea, and asked a woman who lived near the shore whether she had noticed it too. 'Oh no,' was the surprised answer, 'we never look at the sea.' At that time I had a priggish contempt for her insensitiveness: year in, year out, that wonderful sea was at her door, and she never looked at it. Now I know better. She was not there for recreation, but to do her work. Through long familiarity she had ceased to be conscious of it; yet the sea was there, an essential part of her life. To me it was a glorious accident.

The peasant in contemplation of his sheep, wondering how much they will fetch at tomorrow's market, is, we assume, blind to the setting sun which turns the rim of their fleeces to a golden glory. Are we quite sure? And how many thoughts would the man of letters give to a beautiful sunset if he were not certain of the next meal? Not that I am laughing at literary people – far from it; for here am I writing myself, and my dearest ambition, beside the ownership of cattle innumerable, is to write well. But writers must have leisure and freedom from anxiety, which means that the necessaries of life must be secure. Hence literature is really a luxury, and literary people mere parasites on the producers of food. For this reason, in primitive societies poets were generally blind or lame. It is only in a highly civilised and therefore strictly unnatural community that an able-bodied man is allowed to indulge in literature. There is of course no reason on earth for rejecting these luxuries, if we can afford them, any more than we need follow that over-logical Quaker John Woolman, who would not wear dyed wool because dyeing is the denial of natural and apostolic simplicity. But it is well to remember that the plough and not the pen is the ultimate power. If Virgil, instead of going to Rome and writing the *Georgics*, had stayed at home and lived them, the world would certainly have been poorer in poetry. But had not the father made money in agriculture, the son would not have had his opportunity. And in time of stress, which would we welcome most, the young man with his poem on farming, or the old one with a waggon-load of corn?

The true countryman is he who, in face of the various

distractions of modern life, is content to live in the place of his fathers and cultivate the soil. He is the heir of a splendid and continuous tradition; he knows the people as no outsider can ever know them. There is a natural dignity about him, a spacious serenity. Compared with him, the amateur immigrant country-man is like a newly created commercial peer in the presence of a member of the ancient nobility. This man alone, I think, has a real home.

Now at first sight this seems a foolish statement. Real homes, you say, may be counted by the million. Perhaps. But what *is* a real home? The definition of it might puzzle the shrewdest coun-sel in the divorce court, since home is more an atmosphere than a place. In one sense, it is anywhere where two or more people who love one another lead a common life. This definition may fit a one-roomed tenement in Bermondsey as well as an ancestral farmhouse in Dorset. In another sense, home is a place where you can rear a family. But unfortunately families can be reared in all kinds of horrible and unhomelike places, even in a Soviet nursery school. But if we consider it closely, we shall find that a home, to be a real home and not merely an abode, must have continuity. It must provide a fixed and permanent centre for the changing lives that revolve about it, an abiding place which, if we are forced to leave it, we know will still be there when we return. This defini-tion will include the Hebridean black house while it excludes the most luxurious hotels, boarding-houses, and furnished flats. It must also have space about it. Without space there is neither privacy nor individuality. The most fanatical admirer of ants would hesitate to call an ant-heap 'home': indeed the whole ant econ-omy looks like a divine parody of city life, made to be an entertainment for angels and an awful warning to men. From this point of view a flat in a block of flats, however personally furnished and continuously occupied, is not a real home, and a semi-detached house is less homelike than a detached one.

A real home must include people of both sexes and various ages. There is something harsh and unkempt about a house with only men in it, and something insipid and finicking about one with only women. A house without youngsters is stagnant and dreary, while one without older people is unrestful. It must also include animals. The ruler is seen at his best when all his wisdom, patience, and kindness are being displayed in the management of less developed subjects: so it should be with us among our animal dependants. Every child should grow up with animals, if possible with a hierarchy of animals, each in its own place. For there must be no confounding of species. To treat a dog like a child is as bad as to treat a child like a dog. Benighted Victorians used to think that class distinctions were ordained of God; but a visit to Australia, where there is only one class, made me wonder if they were not right. But whatever may be said for equality between man and man, there certainly is none between man and beast; and the community in which the animals do as they please, as they were supposed to do in democratic Athens, will come to no good. By all means disinherit your jazzing nephew, and leave all your substance to a cats' home. But don't let the dogs sit on your arm chairs or the heifers devour your underclothes on the bleach. There is something subversive about that.

The home should be as far as possible self-supporting. In these days of rapid transport it is only in a few remote places that this can be done naturally. Provided he keeps within the law, a crank can do anything anywhere; but unless it is natural, and in no way crankish, the self-supporting mode of life is a valueless affectation. The popularity of allotments shows how many people take pride and pleasure in producing their own vegetables. How much more pride and pleasure they would take in doing the thing more completely – getting porridge from their own oats, milk from their own cows, eggs from their own hens, wool from their own sheep! It adds a zest and interest to life if we know where all the things we eat and use come from, and better still if we have assisted in their production, and not merely taken part in the final and shortest stage – consumption. 'Now, Miss Margaret,' said an old cook reprovingly when I asked her to make my favourite pudding, 'it takes me two hours to make that, and you eat it up in five minutes!' Not that food production has not its sharper side. All unconscious of their doom the little victims play, and we walk among the cockerels and wethers in a Lycidas mood, reflecting that it is not chivalry that makes us spare the females, but the desire for more flesh to the pot. In Australia they used to kill old cows and serve them up at table without long hanging, for the weather was very hot. The head of the house laid down his knife and fork with a sigh. 'Oh dear!' he said, 'Minnie again!' To this day I do not know whether the sigh was for Minnie's toughness or for her untimely end.

By this time the reader, if he has had the patience to get as far, will have begun to smell a rat. The best home is the farmhouse. Oh that my own childhood had been spent there, instead of in the red-brick glare of a suburban semi-detached villa! But perhaps if it had I might now be singing the charms of Tooting. The farmer is a lucky fellow, if only he knew it, and in nothing more to be envied than in his ability to provide such a home, and provide it soon. For the professional man, a wife and children are liabilities; for the farmer they are assets, and early marriage one of the conditions of success. But the born countryman is fast dying out. The town, exhausting the generations as they come, is always crying out for fresh blood, and always getting it. The villages are drained of young life; the brightest boys get bitten with motors or wireless, and off they go, leaving the world's prime industry to fools and sluggards. The girls who should be rearing healthy children to tend the herds are sitting in silk jumpers at telephone exchanges and cashiers' desks. Meantime we get alarmed, and try to revive country life by bringing to the villages the amenities of the town. But it is surely sounder policy to stimulate interest in the amenities of the country, which are always at hand to be enjoyed for nothing, than to import new and artificial interests, which far from keeping the young people at home, will only induce them to go away. For the boy or girl who develops a taste for the pictures will not long be satisfied with the display at the village institute, but will never be content till he can visit the Commodore at Hammersmith every night.

But in the ebb and flow of human affairs there is a law of compensation at work. As the country is depleted of her native inhabitants, their place is taken by a new race of countrymen, those who have tried the city and found it wanting. The amateur countryman is a product of this age, in conscious revolt against the urbanisation of our land and our lives. But before speaking of him in more detail, I want to say a word about the tourist. We often gird at him; but like the casual stranger at a religious service, he is often at the first stage of a progress to something better. Just as an inland boy with a natural taste for the sea may develop it by taking a trip on a penny steamer, or a studious child become a scholar because of the impression made by a school excursion to Oxford, so a man may go down to the country in a charabanc and only return from it in the farm-waggon that serves him for a hearse. The tourist, in his eager receptive mood, is open to impressions that may lead him in the end to some new way of life. Reared beside sluggish streams fringed with elms and pollard willows, I came to care exclusively for open hill, running water, and above all things the sea. This is not altogether perversity and reaction, since, being a Wessex woman, I may find myself as much at home on windswept Egdon as in the lush water-meadows of the vales. But the seeds of this passion were sown at sixteen, when I was taken on a voyage to Shetland to recuperate from an illness. More than health was found there. The physical impression, blurred by other scenes of the same kind viewed later and with more discriminating eyes, has with the passing of years lost some of its vividness. But the spiritual impression is as clear as ever, only, as it is rather difficult to describe, I shall leave it alone, and merely say that if I had never gone on a tourist's trip to Shetland I should probably not be writing here today.

But the tourist, for all his eagerness and sensibility, looks at things from a distorted angle. To him, all the world is a spectacle. But Nature is not to be contemplated only, but to be worked with; not a mere picture, or even a background for human activities, but a living power conditioning our lives, and giving the bread that feeds, the fire that warms, the lightning that kills. The tourist attitude ignores the fundamental truth that behind the beauty of nature is tragedy, and often senseless cruelty. Even in Britain, where everything is keyed down to temperate tones, and there is no place for the obvious horrors of the jungle, I have lain awake on a dark stifling night, listening to the munching of millions of caterpillars, who were systematically stripping an oak wood. Take the sea. We have an immense coast-line, and most of us have made excursions to the seaside. Now in some way the seaside is not the sea. At the seaside you bathe when it is calm and admire the waves when it is rough. You may feel pleasure, interest, and even a thrill of apprehension when a big breaker tumbles you off your feet, or of anger when a stealthy high tide washes away your clothes. But the sea is different – as different from the seaside as a tiger from a domestic cat. No one truly knows or loves the sea who has not feared its fury and distrusted its smiles. It holds the bones of more men than ever admired its foamings from the shore. Long ago, when I was staying in Cornwall, a boy from St Ives came down to my cottage with fish. A heavy sea was crashing on the rocks below, and I admired the beautiful spectacle. The boy looked grave. 'If your people were fishermen', he said, 'you wouldn't say that. Our boats are out today.' Never was snub better earned. I felt as fatuous as the bride, who, when the honeymoon car was jammed on the edge of a precipice, cooed to her distracted husband: 'At all events, Harry darling, you can still kiss me!'

As for trippers, the kind that go shrieking and tooting through solemn glens, leaving a trail of broken bottles and cigarette boxes, we will not speak of them. They are subnormal, only fit for some amusement park a few degrees removed from Bedlam. But the genuine tourist is no more to be despised or jeered at than the genuine enthusiast for popular education. He is limited not by capacity but by opportunity. Most of them work hard in abominable places like Wigan; most of them don't want to do it, but they must, because under present economic conditions there is little employment to be had in the country. We cannot grudge them their few weeks or even few days annually out of town, without which refreshment they would not be able to carry on.

There are a few people who genuinely hate the country in every way and at every season. But very few: for the ancestors of all of us were once countrymen, and the most sophisticated town-dweller must have somewhere about him a touch of the rustic. Even the person who most likes to walk down Fleet Street and look at Man will sometimes prefer to walk in the fields and listen to the larks, always provided that when he wearies of the rural symphony he can switch on to Piccadilly Circus instead. But luckily for those of us who enjoy solitude, the large majority like the country only in summer, when the weather is fine, the days long, and the roads free from mud. Some would rather gaze upon it from the windows of a touring car; but they might as well see a country film at the pictures. There is no real difference, except that in the first case the spectator moves, in the second the spectacle.

To discover the genuine lover of the country, the best test question is 'Do you like it in winter?' or better still, '*All* winter?' For many of the most sophisticated enjoy a Christmas house party. Other questions would be: 'Do you dislike wet feet, mud, the smell of dung, the crowing of cocks, a succession of long quiet evenings, each one exactly like the last? Are you afraid of meeting cows in a narrow place, walking in the dark, crossing a slippery footbridge? Is gas (physical and spiritual) necessary to your happiness?' If the answer to these and similar questions is yes, the candidate may be rejected. He probably takes a dose of rural peace as the Romans took emetics, in order to prolong the city banquet beyond its natural limit.

When I, at that time a pale intellectual, first proposed to work on a farm, I was greeted with hoots of derision. 'Pooh, my dear, you are mad. You'll never stick it. You don't know what it's like!'

To show that I meant business, I went to a small hard-working dairy farm in the month of January. So dark were the nights, so early the hours, that when I was knocked up I never knew whether it was for some emergency or for the morning milking. Coming in with my pails across the muddy yard all stiff with frost, I saw the starlight paling to sunrise long after the working day had begun. Rarely was life more enjoyed. We were fed mostly on bacon and tea: I hate both, but it might have been a Guildhall banquet. I was hungry.

To live in the country without a stake in it is futile – that way boredom lies, and even suicide, for chronic boredom is a living death. Many city people, especially women who have passed the marriageable age and are losing interest in jobs that can never have been very absorbing, long for a cottage in the country. But believe me, to live in the country all the year round without any work vitally connected with it is a hopeless experiment. Without the variety of seasonal labour, its hopes and fears, the companion-ship of your dumb beasts (dumb, thank God, so that in times of stress they cannot answer back or ask silly questions), you will be a prey to the monotony and loneliness that can turn your rustic retreat into a melancholy hell. Without the contrast of a busy day, those long evenings, even when vocal with radio, will be tedious to tears; your companions, if any, will get on your nerves; books will lose their charm and meditation its savour. The most dismal Russian novels and plays, and there is nothing under heaven more dismal, are mostly staged in the country, among people who, because of the number of their servants and the supposed compe-tence of their bailiffs, have nothing to do but analyse each other's dreary souls and make illicit love, not from passion but for want of occupation. Of course there is golf and bridge. Now I am no golfer, and the worst conceivable player of bridge, so that my opinions can have little value. But it does seem to me that these games are recreations, not occupations; and that there can be no recreation where there is no work to be recreated from. I suppose that games of skill and intelligence must be taken seriously if they are to be pursued at all: but is it necessary to play them as if our fortunes in this world and the next depended on the issue of a round of golf? And must I walk the world without a friend because I trumped a partner's ace? Of all the hard faces that are worn by boarding-house spinsters, the hardest are seen on those who play golf all day and bridge all night. Work and some vital interest would make them look and feel years younger.

From these serious considerations I pass to something that seems rather trivial, though actually of great importance. To be happy in the country, you must wear the right clothes. The blethering of fashion papers has thrown discredit upon clothes as a serious interest; but it is no mere chattering females, but philosophers like Carlyle and artists like Mr Eric Gill who have the most to say about them. Next to our bodies, clothes are the most personal things about us, the first and easiest mode of self-expression. That is why the sense of being unsuitably dressed can cause us such acute misery. The pangs of a guilty conscience are nothing compared with the agony of discovering that our stockings are coming down in the Strand. The civilised person's worst nightmare is that he is stark naked in some public place; no doubt the cannibal who has dined too well dreams of sitting under a palm tree swathed in innumerable flannel petticoats. And quite apart from these psychological considerations, which are really irrelevant to the main business of life (for after all, what difference does it make whether our stockings come down or stay up?), clothes may be matters of life and death, as any traveller can tell you. It is vitally important to avoid wearing muslins when rounding Cape Horn, or furs when crossing the Line. You may owe your life to an oilskin or a topee. Why then should the subject of clothes be considered unworthy of serious notice?

By country clothes I do not necessarily mean the specialities of Fortnum & Mason, nor the tweeds, tartans, and brogues decreed by fashion for Highland wear, but what is suited to country pursuits. Thus, though correct Highland clothes are fairly well suited to the climate of the country and the occupations of those who wear them, there are a few scorching days when they are supremely uncomfortable and should be discarded. I remember once appearing at a big house in a cotton frock when everyone else was buttoned up in Harris tweeds, and they looked at me as if I were the Man who revoked at the Portland Club. The only thing to do is to cast away all self-consciousness, and dress naturally and comfortably. This is really more difficult in the country, because everyone looks at you, even the sheep. Above all things don't dress according to theory, conventional or crankish. If in summer my stockings are discarded, it is not (heaven help me!) for aesthetic reasons, since there is no beauty in my legs, nor yet for reasons of health, but because I enjoy the brush of wet grass against bare shins, and passionately hate darning. It is also a charming economy. I wear a kilt not because I am fanatically attached to all things Highland, but for comfort and because I hate the bulginess of female breeches. I wear sandals not in emulation of the Greeks who were beautiful or of Mr Gandhi who is not, but because sandals, like the old Highland brogues, let out the water, allowing your feet to dry as soon as you are clear of the bog. Willie the Scout, who was giving us a hand with the hay, turned up one morning in a pair of new-looking shoes with long gashes on the outer side of the uppers, just above their junction with the soles. He explained that he had cut them on purpose, not only to let the water out, but also to let it in, because if his feet got too dry they gave him pain! So that there may be more sense in sandals than is generally suspected.

Self-consciousness, whether about clothes or anything else, is the curse of the amateur countryman. Like other converts, he does everything, down to watering the horse, with the sense of having a mission. Perhaps it is inevitable. He has given up a good job in the city to keep bees or work as a carpenter. He has constantly to justify himself to disapproving or mocking friends who say: 'I can't think why poor old Joe has buried himself in that hole. What a waste of talents – with his wife and children too!' And he, instead of asking them why the deuce *they* are burying themselves, with their wives and children, in smoke and stench, hastens to explain to them in pamphlets what Rousseau and St. Francis thought about the simple life. The friends remain unconverted, while Joe has lost much time which might have been spent in producing more honey or better chairs. Of course the amateur countryman, who is often intellectual, needs some outlet for his natural activity, and writing will provide him with this; but he will be wise to avoid much special pleading. If he must preach, let him do it sparingly and with a light touch. The best way of converting your friends is to make your home unconventional and attractive, and then invite them to help you with your jobs. Those who come to jeer may possibly stay to work. Or they may not.

To be as self-supporting as possible is an excellent thing; but it cannot be driven to its logical conclusion. Logic belongs to the realm of pure mathematics, and is not much use in practical life – certainly not when you are dealing with the Highland climate or the Highland temperament. Women are laughed at for being illogical; but if they had more logic in them they could never run a home. For domestic life is not a syllogism, but one damned thing after another, and the only causality to be found in it is of the annoying variety, as for instance that if you forget to stir the porridge it will burn. But this is not logic; it is common sense or, if you happen to be in a rage, the devil. It is a good thing to spin and weave your own clothes, but it should be treated naturally, as if you were doing something quite ordinary, like knitting or dress-making, and not a special work of piety that will take you straight to heaven, like performing at a folk dance festival in the Albert Hall. When we gird at factory products, which in general we are quite justified in doing, for many of them are as superfluous and altogether abominable as custard powder, we must remember that even we ourselves, as things are at present, could not live by domestic industry alone, quite apart from the huge population that our industrial development has produced and is obliged to maintain. The hand-weaver may lead the way by setting up a standard of individual work for the mass-producer to copy; but beautiful things can be made in factories, and ugly things on hand looms. One of the ugliest articles that ever vexed my sight was a scarf woven on a hand loom at a charitable institution, to be bought with groans by long-suffering supporters of the charity. No: the strongest argument for producing our own necessaries is that it gives increased interest in work, greater independence of external circumstances and the services of others, and makes possible that neighbourly co-operation in labour that is one of the pleasantest things in life.

The copybook moralist will tell you that money cannot buy happiness; and in spite of all the epigrammatical young men, he is usually right. But it can buy freedom, and in this lies its chief value. When it has passed the level at which it sets us free from anxiety, it becomes merely a nuisance. In one sense you cannot have too much money, for what remains after providing for our daily wants can be given away. But we can have too much of it invested in possessions which are only a worry, like sleeping with the Koh-i-noor diamond under one's pillow. If the tinker's tent blows down in a gale, he need only say damn, set it up again, and dismiss the affair from his mind till next time. But if the millionaire's mansion, full of priceless furniture and china, is wrecked by a typhoon, even if it is fully insured, will he not listen anxiously whenever winds blow loud? To multiply possessions is merely to multiply anxiety, while the pleasure to be had from these things is very small, except to people of the magpie or collector type, who are emotionally subnormal. To own much property is to worry perpetually about flood, fire, burglars, damp, mice, moths, beetles: to spend long hours poring over income-tax returns, insurance policies, Stock Exchange reports, and forecasts of next year's budget. Servants may indeed (if and when they choose) do your dirty work, but they bring an uncertainty and irritation into life which makes the fate of Damocles seem enviable. From the standpoint of mere happiness, St Francis of Assisi was not merely a saint, but a master of worldly wisdom. Ultimately, there are just two things worth having – love and freedom. The first cannot be bought at all, the second only at a price which must not be too high.

9

May: The Long Days Begin

MAY is a month of savage contrasts. Biting winds with hail and snow, and close upon its heels the scorching sunshine of a premature heat-wave; the noisy rattle of a thunder-squall, and then the lingering pensive twilight of the northern summer. It is also a month of great and various charm, richer in colour than any but October, and better able to display it because of the longer days and almost invariably finer weather. This year, the 1st of May was one of the worst days we have had to endure, and the month as a whole was unseasonably wet and cold; but such an experience is exceptional, and in the normal course of things May in the Highlands is as elsewhere the Merry Month, especially the last fortnight, when the lambing is over, the cattle out at grass, and we can take a spell of idleness before tackling the prolonged labour of hay-time and harvest.

Not that we were really idle. Three cows calved; there was milk to handle and calves to feed. The Belties were away at the far end of the farm and had to be looked at daily. There were fences to mend, trespassing sheep and cattle to harry, garden seeds to sow, and the corn to roll. This last was a weary job, as Dick had the strongest objection to pulling any agricultural implement alone, so that it was necessary for one of us to walk beside him with a stick while the other guided the roller. He was a most intelligent beast; in fact he knew far more than was suitable to his condition. Having come to us from a farm with standard hours, he knew when it was five o'clock, and after that time would refuse to put his back into anything. This was annoying, for with our various routine jobs we were rarely free to start work in the field before ten, and therefore had no intention of unyoking at five. The convenience of working your own farm without hired labour is immense. You can go out and come in when you please, without having to curtail work on fine days or invent it on wet ones. In summer we did not milk till half past seven, which left a long afternoon for field work; tea was ready some time between half past six and seven, and we took in the cows immediately afterwards. But in the matter of hours as in all else, Dick had the trade-union mentality, and was the only creature on the farm that showed an unwilling spirit.

May was the crowning month of the bull's activity. Almost every day someone brought a cow. On one occasion the last of these had to be turned away. George was getting six pounds of linseed cake daily, and with the young grass was in fine trim; he seemed likely to repay quite soon the capital invested in him. One week he earned nearly £3, and this money, according to Flora, was as easily gained as when the dentist pulls out your tooth and charges five shillings. Many of George's clients came from her village, and she was careful to see that none of them deserted us in favour of strange bulls in other places. One Sunday, when she was on her way home, she noticed that one of the village cows was bulling. She rushed to the owners and informed them that their cow had come for the bull, and they must take her at once to Achnabo, or it would be too late. The man was at church, but his wife found the cow and escorted her to the farm, where our Sunday nap was, as so often, interrupted by a demand for the bull – one of the few jobs of work one is allowed to do on the Sabbath.

We often found time for a siesta in the woods after dinner, making up for this idleness by a spell of work after milking-time. The brilliance of the young green was beyond belief. In the far north, May has all the youth and freshness of an English April, combined with the long daylight and powerful sunshine of the month before midsummer. The oaks were still bare, but the leaves of the birch were half opened, veiling the branches in a mist of tenderest green; among the dark evergreen pines were vivid cones of larch, so fiercely coloured that against the inky background of a hail-cloud they seemed almost red. Willows of different varieties were gay with softly swaying catkins; the ground was starred with violets and wood anemones and the big primroses that grow in the spongy moss beside the burn. Young shoots of bracken were forcing their way through the carpet of last year's leaves, unfurling softly furred and juicy fronds like bishop's croziers. Birds were singing in every tree; even on small islands in the loch the cuckoo shouted all day long. Along the burns, at the roadside and on the railway embankment, the whins flamed in the full glory of their bloom.

Whin or furze is one of the commonest of bushes in the wilder parts of Britain, but there is none to excel it in beauty. The blossoms are of so rich and vital a yellow that all other shades of this colour seem pale and anaemic in comparison, and their nutty fragrance is far more seductive than the scent of roses. The rose has its thorns, but there are not enough of them to prevent us from burying our noses in the flower. But heaven help the lunatic who would bury his nose in a cluster of whins! This guarded aloofness only makes the blooms with their intoxicating perfume more fiercely alluring. In Cornwall, where first I saw the blaze of whins against a deep-blue sea, there must be more than one variety, for there are blossoms to be seen in almost every month of the year. In Brittany there is a saying that the only time unsuitable for courtship is when the whins are not in flower. Here they are less prodigal; a few blooms appear towards the end of December, after which they gradually increase in number until in April and May the whole place is on fire. From the end of May onwards they begin to fade and gradually disappear.

Besides delighting the eye, whins have their practical uses. The younger and more rounded bushes can be spread with washing without danger of its blowing away. The older ones develop much dead wood, which when dry makes a quicker and hotter fire than anything except perhaps the weathered twigs of heather.

It will boil a kettle faster than any gasring, but needs constant renewing, so that the tending of a fire of whin-sticks is a whole-time job. A whin thicket affords the best of shelter to birds and beasts. Even in the coldest winter a mysterious warmth is stored in its recesses, a warmth that makes itself felt instantly, as though for a moment one had walked backwards into summer. The ground beneath is dry, carpeted with a fine dust of dead twigs and needles, and in July, lovely with slender spikes of foxglove. Here rabbits hold their twilight parliaments, and out-wintering cattle find a lair as warm as any byre; here truant hens may scratch for grubs and, over-bold, end in some wildcat's larder. The glossy green spines make a handy frame for spiders' webs, which, starred with dew, sparkle more brightly than all the diamonds of Africa. On autumn nights, frail threads of gossamer float from bush to bush, and brush against your face like ghostly hands. Speaking of ghosts, Jimmy's eldest son, who was cycling home at night along a road bordered with whins, had his cap lifted from his head by some mysterious agency. An appalling experience. But it turned out that some friends of his, knowing his dread of the supernatural, had stretched a cotton thread cap-high across the road.

The Belties were barely two years old, and none of them had yet been officially served. However we had for some time suspected that one of them was in calf, having no doubt been visited last summer by the government bull from the village, who would often leave his own harem and walk the world in search of fresher charms. By the end of April the matter was beyond all doubt; and in the third week of May a woolly black calf with a white bull's-eye on one side was born in the field. The mother had a great flow of milk, far more than a very young calf required; so every third day or so we took her into the loose-box and drew off the surplus. Though she was a heifer, and of a breed unaccustomed to handling, she stood like a rock without being tied, believing perhaps that the milker was some new and curious variety of calf. At the beginning of the season we had agreed to name any bull calves after the members of Captain Scott's polar party.

Scott himself was born to Dorcas, the minister's cow, but we exchanged him for a heifer of the same age belonging to Flora's people. The Beltie's calf, if a male, would have been called Wilson, but as it was a female, it received the name of Oriana Souper, the lady whom Dr Wilson married. Oddly enough, of the three bull calves born on the farm in 1934, not one is left. Scott, it is true, is still alive, but he was exchanged when a few days old. Wilson, born of Brigid in July, died a week later, and Bowers, Strathascaig's calf, lived for a bare thirty hours. The twelve calves of the season were all females, seven home-born ones, four pure-bred Guernseys from Cornwall, and one Guernsey cross from Strathascaig.

May is a month of great fluctuations of temperature, and we had more than one oppressive day with thunder rumbling in the distance. I am afraid of thunder, and do not mind admitting it. I am also afraid of climbing ladders, but this is a craven fear, and I never confess it to anyone whose respect I am trying to win. But no one need be ashamed of fearing thunder; better men than you or I have been afraid of it. Not because of its danger, for it is far safer to sit under a solitary tree in a thunderstorm than to cross the street. But lightning is terrible, while my neighbour's car, which may at any moment knock me into eternity, is not terrible. It is merely a confounded nuisance: and so I damn it, but do not fear it.

Fear of thunder first roused my interest in the weather. At the age of ten I read the meteorological reports and forecasts in *The Times*, and even kept a record of my own observations. The motive was not scientific but purely practical: I wanted to know at what time during the sultry Thames Valley summer I might be free from that oppressive fear. I learnt by experience and observation the typical thundery conditions of wind and sky, and from this developed a general weather sense, so that my friends would often ask me whether they should take their waterproofs away for the week-end.

Everyone dreads those crawling minutes of apprehension in

the dentist's waiting-room, crueller by far than anything suffered in the torture chamber itself. In the same way, the noisy violence of an actual storm is less terrible than the breathless hush and gathering gloom that come before. Most vivid among the memories of childish terrors are certain black stifling nights in August, when I would lie in bed rigid with fear, my eyes shut tight beneath too numerous bedclothes, my ears strained to catch the first mutter of doom. I still have unkind thoughts of the principal of a hostel for women students, who had the lights turned off at half past ten, and after that allowed no candles, so that I passed a night of storm without the faintest ray of human comfort. In those days I could have spent a fortune in night-lights, for the English Midlands are full of thunder. I came to hate the late summer. The rivers ran low, revealing tracts of rank mud and slimy twisted roots of willow. Their sluggish current was choked with oozy trails of weed, and overhung with drooping vegetation. The dull blowsy foliage of the elms concealed their structural beauty, making them look like giant cabbages. Not a leaf stirred, not a bird sang. There was a ceaseless buzz of insects, and far away the whirr of a reaper; the farmer was gambling on a continuation of the drought. Behind these sounds was the continuous roll of very distant thunder, a trembling of the air, to be felt rather than heard, faint, imperceptible perhaps to heedless ears, but to mine, most horribly menacing. That low, dreadful mutter seemed to threaten my whole existence, and I was sick with fear, a fear more penetrating and fundamental than any other I have felt, even in the hideous uproar of an air-raid. The sky was not visibly overcast, but drained of all colour, and the sun though still shining had lost the clearness of its outline. A livid light was cast upon the scene, giving the trees a flat metallic look, as if cut out in painted tin. The air grew thick and spongy; looming masses of cumulus, copper-rimmed and streaked with ugly wisps of vapour, heaved up from nowhere, drifting like ships becalmed, full of stored electricity. A few big drops plumped into the water and then ceased. The muttering swelled to a prolonged roll; from the opposite quarter of the sky a hot little wind set the leaves rustling and then died away. Complete silence. My hair rose; it was coming at last. But no; the rolling diminished to a muttering, and then ceased altogether. Not just yet. The day wore on until its rayless light was swallowed up in thick and sultry darkness, broken now and then by feeble blinks of sheet lightning. An ominous silence filled the world. Then, long after midnight – crash! The terror was upon us.

This kind of abomination is commonest in the east and midlands, when the slow breaking of a long fine spell produces slight irregularities of pressure, which the weather experts call 'systems of complex shallow depressions'. In the Highlands, thank God, we escape most of this. Partly because in July and August, when thunder is most common in the south, we do not often enjoy a long fine spell to break; and partly because our depressions, though frequent and often complex, are rarely shallow: they move swiftly and noisily along their favourite track between Iceland and the Faroes, bombarding north-western Scotland with the same kind of wind and rain that we get in January, October, or any month at all. Thunder is not often heard; when it comes, it is mostly in connection with hail squalls in autumn or winter. The lightning is fierce and often dangerous; but these storms pass over so quickly that there are rarely more than one or two flashes near the same spot. One wild morning in October the Laird and I were working in one of the sheds at Strathascaig, when I heard a roll of thunder. A black squall was rushing up from the west. Next minute the steading was lost in a whirl of wind and hail. A brilliant flash was followed by a crack of thunder almost overhead. Through the driving storm we saw a mob of frightened cattle galloping up the field for shelter: we ran to let them in. The byre was very dark, and full of a struggling mass of cows and young beasts. Without thinking I turned on the switch. There was a blinding flash: the light in the lamp leapt up in flame and went out with a tinkle of broken glass. The lightning, as we found afterwards, had not actually struck the overhead cable, but must nearly have done so, as the amount of electricity in the air was sufficient to cause a fuse. Even when there is no lightning, these hail showers are heavily charged with electricity, as is shown by the behaviour of a wireless set during a squall. An alarming experience; but I would rather have fifty such than one night of waiting for the mutter of doom.

Thunder (at least for those who fear it) is, like any other dangerous or terrible thing, a great solvent of snobbery. Any port in a storm, and any human company, however foolish and uncongenial, on a dark night with lightning! The emptiest blethering of a bore is music to the ear, if only it drowns that evil mutter below the horizon. Out of the tail of my eye I see a faint prophetic flash. 'Oh, stay!' I say aloud. 'Must you go?' and then add to myself, 'I will gladly listen to all that rubbish, if only I need not be alone with the thunder.' Nothing is lonelier than a human being in a desert place, when first he spies an approaching thunderstorm. One hot day in May I was walking by myself on the hill, far from home, lost in the joy of peace and solitude. Suddenly I heard an ominous moan, and saw behind a distant range a black cloud rising. Would it come my way or not? You can never tell; driven by conflicting air currents, the movements of a thundercloud are quite incalculable. Another moan, a little louder than the first. No use stopping to think: I must get home at once, home to company, talk, and human business. I ran blindly over the boggy flat, splashing in water, stumbling over tussocks. The hill had become horrible to me, a howling waste. Oh for Piccadilly and Leicester Square! Faster and faster I ran, not daring to look behind, lest I should see lightning. After a while it dawned upon me that I had heard no more thunder, and turning my head I saw that the ugly cloud had melted away. I halted, panting. The sweet serenity of the wilderness returned, with added power. Blessed solitude! A whistle sounded. Starting with annoyance, I saw the Laird's shepherd and his dogs crossing the brae in front. I was no longer alone. Damn! Yet if there were another moan from the horizon, I would welcome him, or anyone much less agreeable, with open arms!

The farm was only two miles from the sea, and Peter and I were both devoted to boats and to all things connected with them. Yet for one whole summer we were only once on the loch. A boating expedition takes time, and however little work there appeared to be, we never had a sufficiently continuous spell of leisure to make it worthwhile to go out. Having no car to serve our needs or minister to our pleasure, we were obliged to spend an extra hour in walking to and from the boat. But on the west coast, the sea penetrates so far inland that it seems to exert a powerful influence even on those who never use it and rarely think about it; and no book about this region would be complete without some reference to the loch and what is done upon it.

The loch, though ultimately leading to the open sea, is in appearance completely landlocked. Its entrance is closed by the island of Skye, which sprawls like a starfish between us and the Atlantic. The loch itself is double-faced: its salt and tidal waters, its fish, shells, and weed proclaim it sea, while the unbroken circle of hills and the calm surface in which they are often reflected, suggest an inland lake. In this, the western section, it is four miles wide, and as the nearest hills are nowhere very high, it has not the oppressive gloom of many narrower lochs, and is fairly free from dangerous and sudden squalls. Squalls there are, but as a rule in definitely stormy weather, when boats are not out, and the wide extent of visible sky and water makes it easy to see them coming. The shores abound in sheltered bays, and, provided that the position of certain treacherous rocks is known, boating is safe and simple. But the current, especially in the narrow straits at the ferry, is very swift, and at spring tides there is need of strong rowing and shrewd calculation.

The best general notion of the loch is got by climbing under the face of the crags that overhang its southern shore, until, perched on a ledge 800 feet above, you see its whole extent spread chart-like at your feet. On a clear day at low tide every detail is plain. The bays, promontories, and islands, the rocks above high-water mark and those which appear only for a short time in the twelve hours or even in the fortnight, the various banks and shoals, the limit of the fresh water poured out of the river's mouth, the set of the current, the direction and strength of the wind, the activities of birds and fishes, even the distribution of mussel shells in the shallow inshore waters – all these, and many other things, can be studied at leisure from this aerial perch, and turned to practical account upon the sea itself.

One of the great charms of the loch is the number and diversity of its islands, varying from a flattish expanse of some acres with the ruins of a house and good grazing for sheep, to rocks a few inches above the tide line, crowned with tufts of sea-pink and coarse grass. Some of the larger islands are planted with pines, but wherever Nature is left to herself the only trees, or rather bushes, are stunted birch and rowan. One or two are covered with long heather, which in the old days was sought for thatching. Many have sheltered nooks beside the water with a great abundance of flowers – primroses, wild hyacinth, campion, thrift, and various kinds of saxifrage. The rocks are gay with lichen, smooth and tufted, orange, black, and sage-green. The appearance of all these islands varies with the state of the tide. At the highest springs, their rocky spurs and promontories and the fringe of half-submerged reefs and skerries are lost to sight, and the grassy tops look almost awash. Six hours later, especially when viewed from sea-level, the flat green isle has become a craggy plateau, flanked by jagged fangs, and surrounded by detached weed-covered rocks, each in the occupation of one or two gulls or cormorants. Below the tide line the whole surface is covered with minute marine growths, which cause great variety and beauty of colouring. In late summer and autumn the seaweed takes on a brilliant orange-tawny hue, while the rock with its incrustations is predominantly yellow. Caught by the level rays of a setting sun, outlined and in startling contrast with the deep purple-blue of ruffled water, these isles and skerries have an intensity of colour that must be seen to be believed: the artist who dared to paint them as they are would be accused of improving on Nature.

Perhaps the most interesting of the islands lies close inshore, so close indeed that at spring tides it is not an island at all. Across the narrow strait are a few crofters' houses, and alas for romance! the railway. But in spite of the passing of several trains a day, Heron Island, as I shall call it, is one of the most romantic places on earth. The gaunt and overhanging crags frown down upon it, so near that from their summit a thousand feet above the sea, you seem to have it at your very feet, as if viewed from an aeroplane. The isle is densely wooded with pines, under which nothing much grows but a broad-leaved sedge with blackish tufted flowers. On the landward side is a semicircular harbour, opening from the shallow and rock-strewn channel. On the outer or northern side the land falls sheer to the water in a line of low but precipitous cliffs, fringed with trees to the very edge. At the north-eastern corner is a long natural breakwater, under the lee of which an easy landing can be made on step-like ledges of rock. One of my earliest and most vivid recollections of the loch was of landing at this spot alone in a new boat. The island is a great resort of herons. It was April, and the nesting season was in full swing. Many trees held near their summits large untidy nests, swaying in the wind or with the agitated movements of the birds, who rose in flapping, screaming clouds as soon as I set foot within their sanctuary. On the ground was much debris – broken eggs, dead fledglings, and the fragments of fallen nests. The noise was tremendous. I had just pushed off on the return journey when a new sound was heard, and I saw a long-haired whitish-grey beast scramble down to the water's edge and run out upon the rocks. It turned out to be a billy-goat, which the crofters opposite maroon upon the isle to pick up a living. His habit of pursuing people in boats has given him a bad reputation, but I do not believe he is a savage. Probably he is only asking for a passage home, or for a drink, for I never found any water in the place. But I confess that I did not stop long enough to discover his real intentions.

To get the true flavour of the loch, you must go out as far as possible into the middle. Even here there is no real solitude. The shores of Highland sea lochs are often dotted with crofts, because in many places there is a strip of flattish ground between hills and water which is not difficult to cultivate. You cannot get more than a bare two miles from a house; and though to the townsman this may seem remote, it is by Highland standards a short distance. And the sounds of habitation are more distracting than its appearance. In calm weather, the slightest noise will carry far across the water, especially where there are steep hills to act as sounding-boards. Cocks, dogs, cattle, rowlocks, saws, hammers, carts, and human voices – all these domestic and traditional sounds are heard, with the horrible modern additions of cars, motorcycles, and what is worse than either, outboard engines on the loch itself. These detestable contrivances, which produce the maximum of noise, vibration, and profanity, should be taxed out of existence. But whether I am within earshot of the land or not, there is a point at which my vital connection with it is severed. I am here alone in a boat between sea and sky. The shore where I live and labour is no longer home and workshop; it is a view, a background, something for the sea to break upon and for the hanging up of nets to dry. The gulls that sit exchanging gossip on the skerries may yesterday have followed the plough; but the sea and all its works are far removed from agriculture. It is well known that a sailor or fisherman rarely makes a good farmer. In Homer, one of the commonest epithets for the sea is 'unharvested'. At first sight, when we consider the teeming multiplicity of marine life, and the great and ancient industry of fishing, this seems an ill-chosen adjective. But it contains a profound truth. We are by nature land animals; and in going to sea we are trespassing upon a hostile element, whose modes of life are far removed from ours. We may live and die on the ocean; but our food (for no one would care to eat nothing but fish) is brought from the land. And on land we breed: a sailor may have a wife in every port, but at sea he has no dealings with women. Thus the ocean, with all its danger, discomfort, and privations, of which in these days we have but little direct experience, has been, and I think always will be, a place of refuge, for the renewal of bodily and spiritual life.

May: The Long Days Begin

Even on this landlocked loch, in this ten-foot cockle-shell of a boat, I can feel something of that liberating astringency. There is salt on my lips, the taste of sweat and tears; but life unsalted has no savour. Away from the land, I have no age nor sex, no cares, no passions. Places that teem with vivid memories, places that cannot be passed on shore without a surge of passionate feeling, are viewed from the sea with complete detachment. All discords are resolved, all torments stilled. I am nothing but an eye that watches the wind, arms for the oars, hands to steer or trim the sail, an extension or complement of the boat's own life. For the time being she and I are one flesh, skimming the water like a seabird, served or thwarted by wind and tide, looking for food. For I like to take a handline and some bait. Fishing is one of the best excuses for meditation that exists. Most people are a little ashamed of going out frankly and nakedly to dream. This is not mere hypocrisy or self-consciousness, but the perfectly sound conviction that all able-bodied men and women should be engaged in some kind of activity, of which dreams are a by-product, a kind of phosphorescence.

In the middle of the loch is a bank abounding in haddock which can easily be found by taking bearings from landmarks. A tin of cockles, a long line with fine hooks, an equally long rope with a stone or half a railway 'chair' attached by way of anchor is all the necessary equipment. Row out to the bank, let down your anchor, then your lead, and with the line hooked round your finger sit idly in the stern and wait. If provident and optimistic, you will have opened a number of cockles beforehand. This is done by fitting two shells together by the hinge ends and twisting them sharply. If the haddock are in a taking mood, you will be kept busy. But the sport on this bank is much spoilt by dogfish, who often swallow the bait before the lead touches the bottom. These ugly brutes have the roughest skin imaginable, so that after hauling and removing a dozen or so you will rest your sore hands and abandon yourself to the beauty of the scene.

An hour before sunset and an hour after is the traditional time for cuddy-fishing. This sport is often scorned as childish, but I have rather a fancy for it, probably because it is easy to catch something. Cuddies are pretty fish, about as large as a fair-sized trout, dark mottled grey with white bellies and a black stripe down the middle of the back. They are young coal-fish; the adults grow to a considerable size, and doubtless pose as cod and hake upon the marble slabs of fishmongers. The flesh is quite palatable, though unless smoked it tends to be insipid. They haunt the inshore waters, and on calm evenings can be seen rising in large numbers. The splash and patter made by a shoal of jumping cuddies is one of the characteristic noises of the loch. To catch them, take four or five hazel rods, and bait the hooks with bits of coloured wool. Sit on the rear thwart facing backwards, with the ends of the rods projecting over the stern and the butts secured under your thighs, while your companion rows slowly up and down, so that the gaudy bait is kept trailing on the surface. If the fish are taking, the management of the rods is a full-time job, and a hundred or so can be caught in quite a short time. A more sporting and dignified form of sea fishing with a rod is trolling for lythe. These fish, which are excellent to eat, may be large and lively, and require a certain amount of playing. The best bait is a small eel about three or four inches long, which can be found, but with difficulty caught, for it is unspeakably slippery, in the little pools which form under weed-covered stones below the tide line.

At Strathascaig, in the early part of the year, the Laird and I would be busy with the trammel-net, in which rock cod and occasionally sole and plaice were taken. In the late afternoon we pulled down the boat, and coiled the net, with its triple mesh and row of corks, carefully down on a bag in the stern. The lower corners were weighted with broken railway 'chairs', while a long rope connected the upper end with a petrol can which served as buoy. We chose a spot fairly close inshore, where the net would lie parallel to the beach. The boat was rowed very slowly, while the Laird paid out the net. Next morning we returned and captured the buoy. I rowed slowly, and my companion, straddling on the stern thwart, hauled in the net, yard after dripping yard, clogged with trails of seaweed and twiddling hermit crabs, with here and there a few cod of various sizes, some freshly caught and still alive, others dead and already nibbled by crabs. We rowed home, beached the boat, and stripped the net, the Laird extricating the fish, while I coiled down the empty mesh upon the bag. Gulls screamed round our heads, and prowling cats crept up from nowhere, to see if we intended to gut the fish on the shore, or throw away dogfish. I cannot resist a moment's digression on the fascinating topic of dogfish eggs. This species can have but few enemies, for no single fish seems to carry more than six or seven eggs, as compared with the thousands in every cod roe. The egg, which is the size of a marble, is contained in a hard semi-transparent case which looks like celluloid. It is oblong in shape and at each corner is a long spiral cord like an immense pea tendril. As soon as the egg, all ready cased, emerges from the fish's body, it sinks to the bottom, where it is anchored by the tendrils until ready for hatching.

Ten years ago my boat came over from Skye, nearly new and bright with varnish, a ten-foot dinghy, easily handled, but rather flat in the bottom and broad in the beam. After some years of possession, I decided to fit her with a lugsail. Nothing was bought but a few metal fittings and some yards of unbleached calico. I went to the wood and selected a larch of the right thickness, which having been dead for some time was fairly seasoned. I cut it down, trimmed it, and carried it home, where it was shaped into a mast. I then made a rudder, rough but efficient, and a gaff of a broomstick, fitted with part of the head of an old Dutch hoe, which made it run easily up and down the mast. The sail was stitched with linen thread and bound with rope. Before the wind the boat sailed fast and steered well, but when close-hauled she would make a lot of leeway, even with a ballast of stones in the bow, a fault which was probably due to her tubby shape and lack of keel. Although I generally expected to row up to windward, and have often been reduced to clawing off a rocky promontory with an oar, there was more pleasure in this home-rigged tub than in the smartest and fastest yacht afloat. Passengers, be it noted, were better ballast than stones, for in times of calm or adverse winds they could be made to row, while stones must be thrown overboard, and then, when you wanted to sail again, it was necessary to land on some island to collect more.

The *Cockleshell* was not smart; her paint was worn, her sail, having been spread in Raleigh fashion upon a slippery pier to help my mother disembark, was far from clean. In dry weather she was inclined to leak, and her bottom was often covered with water, smelling a little of seaweed and decomposing cockles. I see her now, running bravely before the wind, with an elegant friend from London on the middle thwart, her well-groomed head outlined against the dirty sail, her pleated skirt drawn close about her knees to avoid the slop of the bilge. She is enjoying it; at intervals she takes out a little mirror and powders her nose.

It is midsummer; in another hour the sun will have begun its brief and shallow dive below the northern horizon. The sky is barred and dappled with opalescent clouds, worked into number-less fantastic patterns – curls, and festoons, and those oddly flattened ovals which weather experts call lenticular, whatever that may mean: to the common man they suggest whales, airships, scythe-stones, flat fish, cigars. Between the clouds are pools of light, from which broad beams shoot obliquely down to mirrored radiances upon the water. Mile upon mile of calm unbroken, now lost in gloomy fiords among the peaks of Skye, then opening out in an ever-widening channel northwards, past Raasay and the lonely isle of Rona, out to the swell and freedom of the Atlantic. The hills look high and far away, their bases blurred with faintly luminous haze. The time of slack water is past, the ebbing tide sets strongly westward, carrying our fancies with it. The anchored boat swings round her bows to meet the current, which, split by the stem, streams past in fan-shaped eddies. The rope, taut and quivering, slants down to where the 'chair', lying with shells and starfish in the aqueous twilight, has suffered a sea-change undreamed of by the railway engineer. The tyranny of shore noises has ceased; there is no sound but the occasional 'chut! chut!' of a blowing porpoise or the shrill cry of an oystercatcher. On a long spit of rock a few inches above the water there sits a solemn conventicle of cormorants. The surface of the loch is as polished glass. Wrecked like an aeroplane upon an ocean flight, a drowning bee floats by; then a huge jelly-fish, purple-pink and oddly marked with trailing frills and fringes. I haul up the anchor, coiling down fathoms of dripping rope in the bows. Settling to the oars I see, half-way across to Raasay, a dark line upon the water, advancing and ever widening. Round the boat the mirrored calm persists, but broken here and there by tiny flutings, faint as the delicate scratches of an etcher's tool. Little flaws of wind, salt with the sea and yet sweet with the bogmyrtle of Skye, herald the coming breeze. The dark line sweeps nearer; it is a belt of dancing water. It gains the boat, and in a moment the still air springs to life. Crisp little waves are hurrying up the loch, leaping and jostling. Before their eager onset the calm recedes, narrows, and vanishes. They tumble over each other, half breaking; they slap against the rocks with spurts of foam, raising a ceaseless babble and jobble; this, I think, and no stormier sound, is what Homer meant by the 'groaning strait'. I ship the oars and haul up the sail; it bellies bravely, and away goes the *Cockleshell*, with a fine hum of the breeze in home-made cordage, and the rush and swing of a following sea. Homeward bound to supper and bed – a fine ending to a good day.

10

June: Work and Play at Midsummer

THE grass had now reached its highest nutritive value, and the whole place was flowing with milk. No one wanted butter, so we proposed to buy in three or four Guernsey heifer calves at about a fortnight old, rear them on our surplus milk, winter them on hay and a little bruised oats and linseed cake, and then sell them as yearlings the following summer. The Guernsey breed is only just beginning to make headway in Scotland, and we hoped that fully acclimatised pure-bred yearlings would be attractive to dairy farmers and command a good price, thus disposing of our milk far more profitably than by churning, even if we could find customers for the butter. I wrote to a farmer near Penzance and arranged for him to send a suitable calf by passenger train, and wire to me when it left Penzance. The telegram came, and we took the trap down to the station to meet the train indicated. No calf was anywhere to be seen. The stationmaster said that he had been warned to expect it at this time; and after an hour or two of telephoning it was discovered that the wretched little creature had missed the right connection at Edinburgh, and was now sitting at Inverness, unable to arrive till next day. I had to arrange for extra feeds of milk, and was wondering how a calf only fourteen days old would stand a journey of more than forty-eight hours. But this particular calf has always been the hardiest and most philosophic of beasts. When we went down again next day, we found her lying in the waiting room, wrapped in a sack, and contemplating the unfamiliar scene with complete detachment. Not a bawl, not a struggle, as we lifted her into the back of the trap and jogged home over the stony roads. On arrival she drank the milk we had drawn for her from the waiting Dorcas, and then settled down with her new companions as if she had never travelled the 700-odd miles that separate Penzance from Achnabo! We called her Dileas (pronounced Jeelas), which is the Gaelic for faithful; she now belongs to Peter, whose father bought her for him as an investment. A little later we got three other Guernsey calves from the same farmer; only one of these arrived by the right train, but all have done well, and the experiment seems to have been worth making.

In June there would occasionally be time to walk for pleasure, and the most frequent scene of our rambles was the hill belonging to Strathascaig. The area of the Laird's property was about 6,000 acres, of which perhaps a hundredth part was enclosed; the rest was open and as Nature made it. Physically and economically, a Highland farm is dominated by the hill, as a ship by the sea she sails. We cannot cross an empty ocean at night without feeling that our precarious lives are infinitely unimportant. More than a thousand years ago, a heathen chief compared the life of man to the flight of a bird through a lighted hall and out again into the dark. Centuries of science and speculation have not taken us much further; and with nothing but an iron plate between us and eternity, the old Northumbrian's parable has lost none of its point. Yet the imminence of the vast unknown makes the well-lighted petty existence of decks and saloons seem not merely tolerable, but even necessary. Two days out from Colombo, at a point just south of the equator, I once watched a liner's cinema entertainment. The films, of the cheap American variety, were shown on deck, with all lights out. Behind the screen, the inky dark was rent with vivid lightning – a continuous play in a clear sky, which made the human show appear not trivial, but rather comforting. So with the hill. Sick of the pettiness of daily life, we go up there to seek escape, and at night return, to find in our fields and homesteads a friendly shelter from the too oppressive sense of infinity.

If I were more orthodox, I should devote the rest of this

chapter to sport. More and more are the Highlands becoming the playground of Britain, and unless the business of sheep-farming and beef production brings in more profit than it does today, the tourist trade will be the only hope, and Scotland be developed as a British Switzerland. If this happens, the fashionable exclusiveness of the Highlands will go: hikers in shorts will eat their sandwiches on the shores of the loch once sacred to Lord A's rod and Viscount Z's gun, and I for one shall not be altogether sorry. Hikers may be and often are objectionable; but at least they are workers who have earned a holiday, and not rich unemployables trying to kill time or work up an appetite for the next meal. But I will pursue this subject no further, for my views are all wrong. When I go fishing I want to catch fish for the pot, and this is easiest achieved by means more proper to the poacher than the sportsman. If I got a chance of poaching, I should certainly take it, and on certain occasions I have smacked my lips over venison unlawfully obtained. But enough of this communistic blether. I am a worker and a woman, and to take sport seriously, one must be comfortably off and a man. Many women shoot and fish, but few of them do it really well, and there seems something a little unseemly and topsy-turvy in this fashion, just as the modern sport of husband-hunting is the product of a wrong balance of the sexes. Strictly speaking, a woman should neither hunt nor kill; one so deeply concerned in giving life should not take it, except in the case of wasps, spiders, fleas, moths, and beetles, where killing is no murder, but an act of charity and mercy. In any case, having no inside knowledge and nothing fresh to say, I must just pass on. The hill is wide enough for every taste. If you do not shoot, you can walk; if you do not carry a rod, you can take a camera; if you have not a camera, you can use your senses alone, and that is best of all.

We hear a good deal about the faculty of observation. In townsmen it is undeveloped, and for obvious reasons. They do not need it in their work; and for them, keenness of sight, hearing, or smell is not merely useless but positively harmful, since it only increases the sum of human misery. In a world of sky signs and pneumatic drills, it is better to be blind and deaf; for to cultivate the senses is the shortest road to misery and death. And as regards observation of human behaviour, the units in a crowd are too numerous and bewildering for individual study. A naturalist observing an ant-hill tends to think of ants in classes and not as individuals. It is the same in town. On a Highland road, we watch a solitary cyclist with passionate interest, wondering who on earth he is and what on earth brings him there. But in a city there are too many to care. We do not even look; we go about our business.

What is this faculty of observation? Obviously not the power of perceiving everything that is presented to the senses, for that is impossible. Even if we had as many eyes as a queen bee, the human brain as we know it would be incapable of co-ordinating all the impressions. The vast majority of our sense impressions must pass unrecorded, like the flight of seconds on the dial of a watch when we are busy. From this incoherent stream we are forced to select; and the keen observer is he who selects the things that matter. For in this business of observation we are not merely the passive recipients of impressions; we are looking for something, and ultimately we see only what we are looking for. That is why the astronomer may fall into a well, or the student of ant-hills be unaware of an approaching storm. These people are not unobservant; they are only intensely selective. Nature seems to dislike over-selection and occasionally takes her revenge.

But Nature, in the impressions which she presents to our senses, is herself a selective artist. The world of vision is in three dimensions, subject to variations of light and shade. Compare a detailed large-scale map in black and white with an equally detailed natural view of hills, fields, and rivers. The first, being in two dimensions, without light and shade, is bewildering in its multiplicity, for every detail has the same value. The second can be grasped with pleasure at a glance, because the play of light and shade selects the details, exalting some and subordinating others. Similarly, the world we hear is in space, so that all sounds vary in distance and intensity. If we could imagine a world in which all noises were equally loud and all came from the same place, what would our sound impressions amount to? A meaningless Babel, such as is heard in an industrial town at five o'clock when all the factory sirens are hooting at once. Some artists are fond of disparaging the realistic treatment of a subject by calling it 'mere photography' – that is to say, a process in which there is no selection, so that everything visible is as far as possible brought into the picture. But this is not an accurate description of the photographic process. It is true that the camera does not select consciously, as the artist does; but it faithfully reproduces Nature's own selection, and that is why the photographer's treatment of landscape is sometimes more pleasing as well as more realistic than the artist's.

All this is in the nature of a digression; the main point is that the observant man will miss nothing that is relevant to his purpose. But the world is full of a number of things, and a good many of them are necessarily beyond the scope even of the most comprehensive purpose. If you are looking for a whale, you may be pardoned for not noticing a jellyfish; if your notion is to get a bagful of trout before dark, you may not stop to look at the sticky hairs of the sundew with a magnifying glass. It all depends upon the centre of your interest. Of course there are people who never observe anything, simply because nothing in heaven or earth has any vital interest for them. But leave them alone – they do not count.

The hill ground belonging to the Laird is all on the southern side of the glen, bounded on the north-west by a line of crags a thousand feet high, which rise almost sheer from sea level. The trough between the crest of the crags and the next ridge contains eight lochs, three of considerable size, lying at different levels and separated by a confusion of rocky hillocks, heathery slopes, and flats of bog and coarse grass. The whole place is not unlike a stormy sea turned to stone. It can be reached by various routes; the most interesting and also the roughest is by way of the crags. These cliffs, being perpendicular and sometimes overhanging, could only be scaled with ropes, but lower down they break into a steep slope dotted with birches and pines. At the point where the trees end and the wall of rock begins, it is not difficult to work westward under the face, and finally reach the top by way of ledges and gullies, where jutting stones and long tough heather afford a foothold. The climber is following a shelf from 500 to 600 feet above the level of the sea, and so directly above it that he can see the changing colours of seaweed and sand, even the iridescent gleam of mussel shells, at the bottom of the clear water. Hidden in the trees below are road and railway; he could throw a stone upon either. The distant views are very beautiful. I cannot say more, because it is impossible to describe them in general terms. The charm of a Highland landscape is the most elusive thing on earth. It depends on a hundred fugitive variations of season, weather, light, even perhaps of the beholder's mood. And when one has seen a particular view very often, it becomes impossible either to generalise or to isolate any one aspect.

And how can we convey to another mind anything so complicated, so hard to seize, as the beauty of a view? In everyday things we have a rough working guarantee that the perceptions of two individuals will agree. When you and I examine together the graining of a piece of wood, or the open pod of a young broad bean, we can be fairly sure that we are both seeing the same forms and the same colours. There is no absolute certainty, but there is a working probability. But when we watch an elusive, complicated spectacle, like a snow shower sweeping across a range of hills, the correspondence between our perceptions seems less certain; and when we come to the emotions that this sight arouses, each individual feels alone, and there is no guarantee that the feelings correspond or can even be communicated. In describing my perceptions to you, I cannot transfer them to your mind: I can only stimulate your imagination to work along similar lines. Suffering, as is well known, is the loneliest thing on earth; so also is pleasure. Not the noisy gregarious sort of pleasure, but the quiet dreamy kind we get on a mountain top. That is why it is better to walk on the hill either alone or with a companion who is silent.

Otherwise the adjustment of two not quite parallel sets of impressions disturbs the harmony.

This wild and spectacular walk along the rock-face must, you think, be very solitary. Perched up aloft, the climber will be far removed from human affairs. Not at all. There is no place on the whole farm, not even in the steading, where one is so obsessed by one's fellow men. The cliff forms a vast sounding board, which collects and magnifies every noise within miles. Some of these noises are natural, but the more insistent of them are human. A train on the line, a car on the road, a hen wife in the township across the loch, oars in the bay, an outboard motor at the haddock bank, voices of crofters and surfacemen, even a clock striking. It is true that great height detaches us from these sounds, so that they seem to come from a world with which we have little concern; but they are there, and it is not until we have climbed the last ridge and dropped into the trough beyond that we escape humanity and enter into the quiet of the hill. It is like turning out of a crowded street into the subdued murmurs of a cathedral. The croak of a raven, the drone of a wild bee, lapping of water on the stones, and the croon of wind in a million wiry stems of heather. Beyond is silence.

As I descend the inner slope of the ridge, I see below me the Black Loch, a long narrow sheet of immensely deep water, with one or two rocky islets on which gulls build. The loch contains rich feeding, and the trout are exceptionally large. They are very shy, and do not take in the daytime unless there is a strong breeze. At the western end is a small round loch at a slightly lower level. The two waters are separated by thirty yards of peaty flat, through which the overflow from the upper loch has cut a deep narrow channel. The lower loch is rimmed with rock, beyond which is more water – but water far below and several miles away. It is the sea between Skye and the mainland, and beyond this sea is the fantastic rampart of the Cuillins. To the east, across the Black Loch, are the hills of central Ross. This neck of land between two waters, where I have spent many hours at midsummer and in

January frosts, is the remotest and most beautiful place I know. Here is a profound sense of boundless space and of eternal rest. Men talk of vast horizons at sea. But the horizons of a mountainous country are much vaster, partly because, being raised up, we can actually see farther, and partly because the gradations of distance and light give the appearance of greater space. Visibility from the deck of a ship does not exceed fifteen miles: we can watch steamers drop below the horizon like flies crawling out of sight upon an inverted bowl. We see smoke from an invisible funnel, which makes us annoyingly conscious of our limitations; there are things going on below the horizon of which we can have no knowledge. No: for the sense of space and eternity, give me the hill. Here is the perfection of solitude. Not of loneliness; for the lonely man, forever seeking a mate, is incomplete, while the solitary man, whether he be as holy as St Jerome or as morose as the Miller of Dee, is always independent – he wants no one. The perfection of loneliness is found in the midst of an indifferent and uncongenial crowd, where the individual counts for so little that even in his own eyes he becomes no bigger than a grain of sand. But solitude is best enjoyed alone with Nature, when the man is monarch of all he surveys; he feels like the only person on earth, a kind of Adam, but one in whom Eve is merged or forgotten. For the wild places bring release as far as it is ever possible from the tyranny of sex. Hill plants and animals have no less vigour and fertility than those of the glens, but they seem chastened by the austerity of their surroundings, so that they do not flaunt themselves so shamelessly or stir the blood of the human observer. The Greeks had good reason for making the goddess of hunting and wild things a virgin.

Many a time have I lain on the hill in June, secure from all wrong and disturbance, baring my skin to the cool wind and hot sun; a gracious contact, warmth without pain, caresses without bitterness, delighting the body and leaving the soul free as blown thistledown. The air is fragrant with bog-myrtle, the ground starred with shy little flowers – potentilla and orchis, wild lobelia and butterwort. Round white clouds move slowly on, trailing their purple shadows over the green and dun of the hills. Small dragonflies, of a bright metallic blue, dart hither and thither in pairs. A wild bee drones on his way. This hum of a passing insect, coming out of nowhere and fading into silence, is the loneliest sound in Nature. Like the noise of a distant train at midnight, it makes us conscious of the mysterious origin and destiny of human life. Absolute peace. Unbroken quiet. Were it not for the unknown and therefore formidable pain and struggle of actual dissolution, it would be good to die now, in a place like this, without any further experience. This is a day of days, one in a thousand: we could not bear too many of them. Like all rare things, it would be cheapened by repetition.

Mostly the hill is in a sterner mood, but not less beautiful for that. The enjoyment of wild and austere scenery, a comparatively modern development, seems to be becoming a necessity of life. 'The new Vale of Tempe', says Thomas Hardy in a famous passage, 'may be a gaunt waste in Thule: human souls may find themselves in closer and closer harmony with external things wearing a sombreness distasteful to our race when it was young. The time seems near, if it has not actually arrived, when the chastened sublimity of a moor, a sea, or a mountain will be all of Nature that is absolutely in keeping with the moods of the more thinking among mankind.' But the pity of it is, that as the need of these things increases, they become more and more difficult to find. As there are hardly any genuine savages left, so there is hardly any wild and unspoilt scenery. Whatever is spared by industry and settlement is overrun by people seeking change and refreshment. One cannot blame them; but where is the lover of wildness and solitude to go? The rocky ice-bound coasts of Labrador, the howling wastes about Cape Horn, even the Great Ice Barrier of the Antarctic, are all in process of being opened up to the tourist. Adelie Land, which has an average wind velocity of about eighty miles an hour, will probably not be a pleasant resort in our lifetime; but the day is not far distant when the lover of solitude will have to content himself with plugging his ears and contemplating the remotest stars through a high-power telescope. The earth is far too small to be alone in.

Reluctantly I leave my resting-place, and drop down towards one of the smaller lochs, which once was a good fishing ground, but is now little better than a swamp. Sixty years ago it was closed and deepened by a dam built up with stones, wooden stakes, and great blocks of turf. But two years ago a terrific rainstorm, which must have been a cloudburst on the hill, made a breach about fifteen feet wide, through which most of the water drained away, leaving a shallow marsh and a welter of debris suggestive of an earthquake. Near this loch are extensive 'flats' – level stretches of wettish ground, covered with moss and coarse sedgy grass. There is nothing pleasanter to walk over barefoot. The soft yielding moss, the tickle of the grass-blades, the bubble of water and warm peaty ooze between the toes, are a continual delight. The heathery slopes, especially where there has been recent burning, are too rough to cross unshod; but sandals and bare legs make pleasant walking in summer. The sandals let out the water as soon as it enters, so that the feet are never stuffily damp.

At a lower level is the long narrow sheet of water called Loch na Gillean, from which issues the burn that generates the Laird's electricity. This burn has carved itself a gravelly gorge, fringed lower down with birch and hazel thickets; on the right bank and higher above the stream is a well-defined path, from which magnificent views can be had to the north and east. About 200 feet below the summit lies a small sheltered flat with a spring of intensely cold water. Here, one fine weekend in September, I decided to camp, taking with me Colin, the Laird's young nephew.

We had been working hard at the hay; but at four o'clock on Saturday we left the Laird and the Scout to carry on without us, and began to pack our equipment. We had exalted ideas of comfort, so that our burdens were immense. On my back was a

Skye creel packed with blankets, hay, and utensils. In my hands I carried bags of food, including milk and clotted cream. Colin had a sackful of bedding, tent, and utensils, an axe, knives, and more food. Thus laden, the climb of nearly 800 feet in the hot sun was fairly stiff; and owing to the unwieldy shape of the creel and the bulk of hay, a stumble was fatal; one could not rise without help. Colin's load was heavier but less cumbersome. Halfway up, at one of our numerous halts, we photographed each other. But we had not much time for such diversions. Arrived at the spring, we pitched our tent at once, and while Colin went off to collect sticks, I made our beds with great care. No freshly cut bracken, with its powerful smell, but a thick layer of twiggy heather, covered with hay, a groundsheet on top, and pillows of flour bags stuffed with more hay. We had plenty of blankets and a hot-water bottle; between the pillows was a candle in an improvised holder, so that we could read ourselves to sleep. A fine fire was soon going, over which Colin began to cook our supper. Every camper, especially if alone or with only one companion, knows the subtle charm of evening in the wilds, when the last light is fading from the hilltops, and the darkness of the earth invades the sky. Then Nature is very near, and the small familiar human comforts, food and fire, candles and hot-water bottle, seem all the dearer in comparison. We sat long by the blaze, talking most unromantically, until, having first cleared the tent of those small active spiders that live by myriads in the heather, we rolled ourselves in our blankets and fell asleep.

To wake at dawn in a closed tent of a green Willesden canvas is like finding yourself in an ice grotto, or under water. A cold greenish diffused light is shed upon everything; faces look ghastly, and the commonest objects seem to have suffered some queer change. It was cold too. Peering out through the ventilating flap, I saw the whole world wrapped in a wet white fog. The hot-water bottle had long lost its warmth; and insulated in our roll of blankets, the amount of heat we derived from one another was very small. Colin was still asleep; watching his mop of dark hair on the ice-green pillow, I wondered if I would wake him: I was hungry, and it was his job to make the porridge for breakfast. I let him sleep on. I began to wish, in the futile way of older people, that he would never grow up any more. At fourteen he was still simple, sexless, natural, in his own way a perfect companion. Next year he would be interested in ties, brilliantine, and girls. In a few more years he would probably be balancing figures in a bank, and taking fluffy damsels to the pictures. Already faint but significant symptoms could be observed: he had lost his taste for books of adventure and horror, and wished us to believe that he could no longer sing treble. Damn. Of course it was inevitable, and brilliantine and girls as natural as scout knives and brown scratched legs. But all the same, damn. I shook him roughly, and pushed him outside to make his toilet at the burn. Left alone, I thought of the dinner we were going to make for the uncle and aunt, who had promised to come up and share our midday meal. Mulligatawny soup out of a packet, bully-beef hash with potatoes and fried onions, followed by boiled semolina pudding with jam and clotted cream. What could be better? Whistling and the crackle of sticks announced that Colin had returned and was getting breakfast. I went down myself to the burn for a wash; and when I came back out of the trees the mist round us lifted like a theatre curtain, revealing a cloudless sky and the whole world sparkling with drops innumerable. Immediately below us, the glen was packed to the brim with layers of fog soft as cotton wool; the higher hills stood out like detached rocks in a foamy sea, while the lower ones were being alternately submerged and exposed by swirls of eddying vapour. Down there, uncle and aunt, farm, cows, hay – all the apparatus of daily life and work, lay smothered in a blanket of dripping fog: the new radiant world of the morning was ours alone. 'Come on,' said Colin, 'the porridge is nearly done. We needn't hurry down, as the hay won't be dry enough to work at for ages.' He gave the black iron pot a final stir, ladled out the contents, and started a fresh song:

Highland Homespun

'In eighteen hundred and sixty-one
The American Railway was begun,
The American Railway was begun,
The great American Railway!'

The weather remained hot and very fine. All the crofters were busy cutting peats, and I felt rather sad that we could not do the same. But there was no good peat on the farm, and such beds as there were had not been worked for years, so that it would have needed much labour to clear them for cutting. In any case, with unlimited supplies of wood at our doors, which could be cut at any season and in any weather, it would obviously not pay to hire Murdo for a job that had little but sentiment to recommend it.

Incredible as it may sound in a country with an annual rainfall of nearly seventy inches, we were beginning to run short of water, and for a few days the gravitation supply, which was gathered in a cistern from various small springs and field drains in the long cornfield and Ladystone, ceased altogether, and we had to carry water for the house from an old well at the bottom of the steading. Except in the peat mosses, the soil is light and pervious, and where it lies on a steep slope, dries out very rapidly, so that two or three weeks without rain will cause a temporary water famine. Luckily the burns and freshwater lochs always provide an abundance for stock, but in June the field and garden crops, and sometimes even the grass, will suffer much from excessive dryness. What would happen if we were to experience the long spells of drought so common in the south, I cannot imagine. But as things are, when twelve or fifteen inches of rain may easily fall in a single month, quick drainage is our one salvation; without it, the land could not be cultivated at all.

We took advantage of the dry weather to attack the weeds in both gardens. The West Highland weed is immortal; so great is its vitality and tenacity that nothing, neither the hardest frost nor the hottest sun, will destroy it. The prevailing dampness of the climate not only stimulates the growth of weeds, but makes it impossible to burn them. In the field garden, the chief pests were sorrel, spurry, chickweed, groundsel, and a vigorous little plant with a small white flower, which seems to have many generations in one summer, but I do not know its name. In the walled garden, we had chiefly docks, hemlock, nettles, and thistles to deal with. At this point there should follow a general commination on weeds. I know a retired fisherman whose life is now one long warfare against them. In our wet and weedy climate, his chief delight is in flawless gravel paths. He will not grow peas, because it is impossible to keep them absolutely clean without uprooting the plants themselves. To him, the loveliest sight in a garden is a bed of well-worked soil with nothing at all in it. Like the ancient Romans, he makes a solitude and calls it peace. As for me, I cannot take the weed crusade very seriously. I know this is wrong-headed and an excuse for laziness, for what is more back-breaking, disagreeable, and unrepaying than an afternoon's weeding? Weeds are, of course, plants in the wrong place. Not from Nature's point of view, or they would not be there at all, but from the gardener's. Their competition spoils his crops. But if the garden plants, so carefully sown, manured, tilled, and tended, cannot hold their own against these wild interlopers, why not let them die? The whole business of civilised man is, it seems, to promote the survival of the unfittest, thus upsetting the balance of Nature. If we had sense, we should prefer the hardy dock and thistle to the delicate cauliflower and celery, which can only be kept alive by coddling. But we have palates; and as I happen to like celery and cauliflower I will say no more of this. Only may somebody else do the weeding.

The potatoes had become incredibly dirty. At the near end of the field the young plants were almost smothered in a forest of spurry, a shocking sight for passers-by, some of whom did not hesitate to make their comments. Peter was due to get his holiday on the 7th, and after his departure Mr Gordon and I got to work with the hoe. I heartily endorse all the complaints ever made by anybody about the miseries of hoeing. Perish the day when I first thought of planting those potatoes! The sun was beating down in the relentless fashion of the north, where, if there is any heat at all, it is very fierce. Clouds of flies buzzed round our heads as we hacked at the obstinate roots, knocking off a good many potato sprouts in the process. It would have been pleasanter to work in the long light evenings, but the midges made that unthinkable. One day I hoed six drills, and, feeling rather proud of my handiwork, boasted of it to the shepherd's wife, who had come down with a basket of eggs for the baker. She was not a Highlander, and frowned in disapproval. 'You should get a man-body to do that,' she said severely, 'it's not right for you at all.' I replied that I could not afford to hire anyone, which was true; but she made me feel that somehow or other I should have been doing something else, and whatever glamour there may have been about hoeing potatoes departed from that hour.

A pleasanter job was the mowing of thistles, nettles, and docks. We were not much troubled with bracken, which, being useful for bedding, was an asset rather than a nuisance. But on most Highland farms with extensive hill grazings, bracken is a very serious pest. Once established, it is very hard to eradicate. It increases rapidly, and if allowed to spread unchecked will ruin large tracts of good pasture. Under its dense shade nothing but hardy weeds or coarse and worthless grass can hold their own; and in the thickest brakes the soil is often quite bare. By frequent and regular cutting it can be weakened and ultimately destroyed; but cutting on a large scale has now become impossibly expensive. The scything of bracken, though it requires no special skill, is a wearisome job. The young stems are soft and succulent, but when mature they soon get tough and stringy, offering a stiff resistance to the scythe and quickly blunting its edge. Many thickets conceal stones, rocky outcrops, and trails of derelict wire. In the eyes of the mower, the only merit of bracken is that it stands up well. But the pest has its uses. In a non-arable country it is cut for litter, and well it serves the purpose; it is more absorbent than straw, and produces manure of better quality. It also makes useful packing material for the protection of potatoes from frost. I confess that I have rather a weakness for bracken. At every stage of its growth it is a thing of beauty which, even for the sake of good husbandry, one would hate to destroy. The tender fronds of youth, soft with silky hairs, curling like croziers, or at a slightly later stage, like small clenched fists; the dark-green spreading leaves of midsummer, the fiery red and gold of autumn, the quiet russet of the dying fern: what could be better? No bracken at all, the farmer would say, and my common sense would agree with him, just as I should feel guilty in admiring a cornfield full of charlock and poppies, or a meadow full of moon-daisies and knapweed. But still, I would not care to see the bracken gone altogether.

With nettles and docks we are on surer ground. These are undoubtedly pests. They are useless, ugly, annoying, and persistent. The nettle has no merits; the only merit of the dock is that its leaf will soothe a nettle-sting. Luckily for us, they often grow together. This little fact must have been overlooked by those who assume that the universe was not made for man's convenience. It is true that during the war nettles were used as a substitute for spinach. But does anyone really like spinach well enough to want a substitute for it? And if so, why not find a substitute that has no sting? It is said that if you grasp a nettle firmly enough you will not be stung, and that it was created to teach us the importance of facing difficulties instead of shrinking from them. But we, with a wholesome pagan dislike of sermons in plants, have avoided our lesson by the use of gardening gloves, clippers, and scythes. On no account fumble with a nettle; but why grasp it firmly when you need not touch it at all? Another unpleasant thing about these weeds is their intimate connection with man. We do not (Heaven forbid!) deliberately grow them, but wherever we go, they follow. They do not flourish in the wild, but delight in ill-kept gardens, dirty fields, useless corners, and that fearful no-man's-land, that combination of midden, rubbish dump, and smithy yard that surrounds almost every Highland cottage. They are perennial as sin, indestructible as folly. They seem to have no diseases, no enemies. Opposite the byre was a triangular plot of ground which years ago was used as a garden, and still contained a number of prolific though sadly neglected blackcurrant bushes. The soil, enriched by drainings from the dung-heap, must have been very fertile, for the bushes were choked by a vast forest of succulent nettles, some of them nearly five feet high. While they remained there was no possibility of picking the fruit, and we planned their destruction. Among the currants there was no room to swing a scythe, so I took a bill-hook and began to slash them down. The place was cleared at last, but I have never been worse stung, and for two days my legs and arms were so swollen and inflamed that life was a misery.

The only satisfactory way to destroy docks is, as everyone knows, to dig them up by the roots and burn them. We had no time for this labour, and needs must content ourselves with cutting them several times in the season, which had the effect of weakening and stunting the plants, though at the same time caus-ing them to go to seed with incredible rapidity. The farm had been much neglected in the past, and the whole steading and nearer fields were overrun with this pest. At Rattray's the docks were even worse, and it was said that the previous tenant had an enemy who used to sow dock seed in his garden to spite him. But so prolific is the dock, that this explanation is hardly required.

On Tuesday, 5 June, Rona calved. This cow had been a daily cross ever since I took her over from the landlord on our first coming to Achnabo. Actually she was an excellent young beast, out of a cross-bred Shorthorn cow by a pedigree Red Poll bull, and the heaviest milker we had. All the neighbours admired her, but I think that few would have cared to possess her, for not merely were her teats as stiff as Aunt Dammit's, but she kicked and bucked like a wild buffalo, and before any milk could be drawn had to be stupefied by the tightening of a rope round her body, which slowed down the movement of the heart, and if carefully used would make her quiet without doing any harm. The first summer Alec contrived to milk her, but it took much time and effort, and whenever there was a thunderstorm I hoped she would be struck by lightning, as all my stock was insured against fire. But before the season ended we had put two calves on to suck; the experiment was successful, and this year I had decided not to try milking her at all, but to let her suckle her own calf and another of the same age.

We took her into the byre, tied her up and put on the calves. They had a hard time of it, for they were young and weak on their legs, and Rona was so furious with the interloper that in her endeavours to get at it she would risk injuring her own offspring. At this stage the calves were satisfied with a bare third of her total yield, and how to get the rest of the milk from her distended udder was a terrible problem. It was quite useless to rope her legs, for they whirled like windmills, and we had to twist the body rope so tight that she was in imminent danger of falling down in a faint. To sit on a stool or milk into a pail was a hazardous affair, but the one-hand method was so slow that Peter risked it, and stripped to the waist succeeded in wringing the rest of the milk from the indignant and protesting Rona. The trouble was that Peter must leave in two days' time; the cow required a man's strength, and none of our male neighbours were able to milk. To cope with her unaided for a fortnight seemed beyond my power, so I ordered a truck and entered her with calf at foot for next week's sale. Meanwhile Flora's brother came twice daily to help me. The calves were now sucking better, and only one teat, of immense size and most diabolically tough, defeated their combined efforts to extract nourishment. The second calf, a Highland cross-bred, was impervious to kicks and buffets, and succeeded in getting a larger share than Rona's own daughter, a dainty little creature whose genteel manners and appearance earned for her the name of Maud. If we could keep going till Peter's return, it would be a pity to sell a profitable beast. I wired to the salesman, countermanded the truck, and kept her. This was not the end of our trouble with Rona, but as the calves grew stronger and hungrier and the volume of her milk declined, she became less difficult to handle, and in the end would allow both calves to suck her in the field.

Peter came home in time to help the Laird with his clipping. He would leave the farm immediately after breakfast, walk over to Strathascaig, stay until five o'clock, and be back at Achnabo for tea. At this time we had the Belties, Pendeen, and the stirks in an extensive no-man's-land on the way to Strathascaig. This tract of nearly fifty acres would, if drained and limed, have provided excellent grazing, and even as it was would summer a fair number of cattle. On paper it was part of our farm, but in actual practice I rarely used it, for the fences were completely derelict, so that it was necessary to herd stock as if on the open hill. In some places the grass was knee-deep, and with so much food at hand, there was some hope that the cattle would stay there. We used to visit them daily, and so far there had been no trouble. So we became a little careless, and on Sunday we left them unvisited. Monday dawned hot and fine. Peter went off to the clipping, and after finishing a few odd jobs about the farm I set forth to look at the Belties. Up and down and to and fro I wandered, seeking them everywhere, but they were nowhere to be found. This was not surprising, for if a sudden fancy seized them there was nothing to prevent them from vanishing in any direction they pleased. But what really puzzled me was the absence of any fresh cattle-tracks at the upper end of no-man's-land. Having calves to feed at midday, I returned home with the intention of looking for them again in the afternoon. Outside the farm gate I met one of our village neighbours, and asked him if he had seen the Belties. He said no, but he had heard that someone had noticed them late on Saturday night at a place a mile or two on the farther side of Strathascaig. What could have taken them there I could not imagine, nor was there any special reason why, having gone so far, they should not go a good deal farther, especially as they had had thirty-six hours' start.

Leaving Flora to attend to the calves, I set forth dinnerless upon my quest. The day was close and still, and before I had covered half the distance to Strathascaig my feet were sore and temper frayed to ribbons. In muddy sections of the road there were confused cattle-tracks, fresh enough to have been made by the truants of Saturday night, but as there were plenty of crofters' cows about on the roads, I could not be sure. I reached the Laird's clipping shed about two o'clock. The place was full of sheep, fleeces, and clippers; at the door stood the Laird, with a cheroot between his lips, blandly surveying the landscape. He saw me panting up the brae, and greeted me with a slow and irritating smile. He wore the stained and tattered kilt that was kept for sheep functions, but had only been sorting fleeces and looked cool and unruffled. In furious tones I asked if anyone had seen my cattle. A voice from inside the shed replied that they had been up the glen on Saturday night, but had been seen returning on Sunday. Where they were now was another question.

Peter left the clipping and joined me in the search. Halfway home we met a neighbour of Jimmy's, who told us that he had met the cattle on Sunday afternoon, and had put them back into no-man's-land from the bottom end, which accounted for my having seen no tracks at the usual entrance above. Had we visited them on Sunday we should have discovered all this. But if they were still in no-man's-land, how had I failed to find them, as it is not easy even for a very stupid person to lose sight of a dozen beasts in fifty acres? But as we pondered this problem, an awful hunger assailed us, for Peter had missed the clippers' dinner and I, as already remarked, had not waited for mine. At that moment we heard the noise of approaching wheels, and as if in answer to unspoken prayers the baker's orange van appeared before us. We stopped it and bought two dozen sugar-buns on tick. Retiring to the shade, we ate the lot.

After this we felt better, and began a thorough search of no-man's-land. At last, deep in a thicket of high rushes, bracken, and fallen pine trees, we saw Pendeen's horns. The rest was easy. But we had to listen to a good many jokes about lost cattle; the Laird used to tell us that someone had seen Dick on the road, or ask whether we had missed Dorcas or the Guernsey calves. That was the longest trek the Belties ever made, and what started them off upon it remains a mystery. They had plenty of grazing at hand, and none of them were bulling. Possibly they did it to teach their owners a lesson, or more probably to make trouble on Sunday. Only we did not look for them on Sunday!

11

Interlude: What's Your Hurry?

IF you want a job of work done in the West, and ask a man whether he can get it finished by a certain time, he raises his eyebrows and says in a slow soft voice, 'What's your hurry?' He may go to explain that tomorrow he must attend a neighbour's funeral, and the next day, if the weather is good, he must give Sandy a hand with his peats or go fishing with Angus; but more often he will just put that searching and unanswerable question, 'What's your hurry?' For after all, what *is* your hurry? If it is a pair of boots you want to have soled, you can always wear the old pair with holes, or if it is too wet for that, sit by the fire and dream until the cobbler is ready. Provided that there is food in the cupboard (and the next meal is possibly the only thing under the sun worth hurrying for) we can afford to wait, and the other fellow's delays will only give us a good excuse for doing nothing ourselves. The author of the book of Genesis clearly regarded work as a curse. Now I am so far bitten with the Sassenach bug of industry that I am inclined to think work a blessing, but measured, leisurely work, except where you are circumventing the weather, as when you rush in a load of hay: but then an element of sport enters into it, which makes all the difference. It is hurry that is the curse. All frenzied activity – all hustling and bustling, scurrying and hurrying, buzzing and jazzing, hooting and tooting, is destructive of peace and beauty, and will in the end degrade rational humanity into a collection of jigging automata in a mechanical world.

Nothing that is great, beautiful, or dignified condescends to hurry. In God's sight a thousand years are as yesterday. The stars, the tides, the seasons, move in slow and solemn procession. Captains and kings ride on stately chargers; they do not scoot on motorcycles. If we saw some venerable personage in the robes of office, some judge or archbishop, scuttling across a street to avoid fast traffic, our respect for him would somehow diminish. At funerals, when we wish to pay respect to the honoured dead, we walk slowly and play slow music. We admire women with slow soft voices and slow dignified gait. We listen with pleasure to speakers and preachers who take their time with us and do not drown us in a torrent of shrill verbiage. Those who hurry are either slaves driven by external necessity, or neurotics goaded by their own complexes, or criminals fleeing from justice, or half-wits trying to amuse themselves. In Nature they have their counterpart in the restless world of insects. The man who said 'Thou sluggard, go to the ant' was a fool. Ants, like bees, are models of industry, but they were not given us as an example but as an awful warning of what efficiency may lead to. There is no sadder sight than the scurrying crowds of clerks and typists who rush with attaché cases to catch the 8.45 to town. How many of them would not really rather be walking slowly behind the plough, or sauntering along sweet-smelling lanes at the tail of a herd of cows?

What use is all this hurry in any case? If, as scientists tell us, the whole universe is running down like a clock, why not run down too? The power that winds it up again will no doubt wind us up also, so why bother? Time sweeps us on in pitiless spate, as a flooded river carries down the leaves; whether we lie still or gyrate wildly, it is of no account, since all roads lead to the same end. As I write, I can see a waterfall tumbling down the hill, carrying thousands of gallons of water to waste in the day. But if it were harnessed to some hydroelectric scheme, it might be lighting our homes, or supplying power to a munitions factory, thus enabling us to blow each other to bits in the next war. Why not leave it as it is? *What's your hurry?*

Now all this is perhaps not practical wisdom. Whether the universe is running down or not, we must get on with our work;

Interlude: What's Your Hurry?

and we cannot work successfully unless at the moment we believe our job to be the most important thing in the world. But it is good to remember that from a more distant standpoint our best work is supremely unimportant, and that to fuss and bustle round it too much is to banish from our lives all premonitions of the ultimate peace. Besides, we need to cultivate patience with the Highland cobbler who keeps us waiting for our shoes. He is lazy and procrastinating, but he can raise a question which it would baffle a syndicate of philosophers to answer. What's your hurry? What *is* our hurry?

I wake with a start, and discover that I have been preaching. Why is the lay pulpit such a temptation? I have not even Stevenson's excuse of Scottish nationality or Calvinistic upbringing. And what is worse, all this justification of sloth has a tang of insincerity, for no one more vigorously damns the Highland *laissez-faire* than myself, that is to say, when it suits me to do it. I hate waiting about, I hate putting off anything I have set my mind to do. So let me drop all this blethering, and try to find out in an impartial sort of way, without anger or resentment, if that is possible, why this planless, timeless dawdling way of life, so charming or so maddening according to the mood you are in and the purposes you cherish, has for its special home and temple the western side of the central watershed. For so it is. As soon as the rivers stop flowing east to the German ocean and start flowing west to the Atlantic, the rhythm of life slows down. We get up late and work more casually. No one blames us if we spend an hour or two leaning against the stable-door-post, lost in conversation or thought, or stand motionless in the middle of the road, contemplating the stars reflected in a puddle. Our faces are turned away from the chill endeavour of morning towards the dreams and gossip of evening. 'Comrade', says A.E. Housman in one of his last poems, 'gaze not on the West.' The warning is no vain one. If you let it, the glamorous setting sun will cast its rays backward beyond noon, until the whole day becomes a long-drawn prelude to bedtime. You cannot watch the evening star above the hills of Skye without feeling that all purposeful labour is a foolish waste of time, an ugly violation of Nature's fundamental peace. Yet beyond the dreamy verge of the western ocean lie no islands of the blest, but the strident inferno of American big business. Perhaps it is all an illusion. Travelling in search of peace is certainly an illusion. '*Qui multo peregrinantur*', says Thomas à Kempis, '*raro sanctificantur.*' Those who gad about are not often the better for it, especially now that modern luxury travel forbids us to exercise even the virtues of patience and endurance. The sacrament of evening is renewed daily, even in Manchester: better to wait for it where you are. But if you want perpetual evening, come to the West. Why has November sunshine a peculiar charm? Because the sun is so low in the sky that the brilliant purple and tawny colouring we mostly see at sunset persists all day. There is the same quality about the West. We never hear the sunset gun too soon.

Let us put these rather subjective considerations aside, and turn to the physical climate. It is very soft, or enervating, as the Puritans would say – enervating like love and warm baths, and many another soft and gracious thing. There are no extremes of heat and cold to set your nerves on edge or stimulate you to rash and feverish action. There are plenty of wet days to drive you from superfluous labour outside to the delights of the fireside arm-chair. In a place like St Andrews, the only way to keep warm is perpetual motion, mental and physical: you must walk or golf all day, and then stand at wind-vexed street corners and argue about sin. Here you may lie all day in the heather, or if wet, in a barn full of hay, murmuring the inspired words of the Psalmist, 'So He giveth His beloved sleep'.

There is no doubt that the softness of the climate has its effect upon the natives. Though often lazy and inefficient at home, they do extremely well abroad, even when all allowance is made for the fact that the most enterprising are always the first to emigrate. But it is likely that the climate has rather developed an inherent weakness than created a new one. It is easy to overstress the characteristics of a race, but it is pretty clear that the feminine softness and pliancy that give the Celt his charm have made him too apt to yield to external pressure. One of the most remarkable things in history is the almost complete extinction of Celtic language and individuality, not by extermination, but by too ready an assimilation of the manners and speech of the conquerors. In the last century before Christ, the north-western half of Europe was Celtic, as much of it is still fundamentally. At the present time, those who can speak a Celtic language are a mere handful on the Atlantic seaboard. There is some weakness here that has nothing to do with the balmy airs of the Gulf Stream.

Whether in his own tongue or another, the Celt talks too much to be industrious. Talk is not only the solace, but the enemy of work. A good talker is rarely a good worker, except where talk is work, as it is deemed to be at the bar, in the pulpit, and in parliament. This is natural, as only a genius could combine work and talk without letting either suffer. If the Highland climate is bad for work, it is certainly good for talk. What are those constant showers for, if not to provide intervals in which we can down tools, and, sitting under the dyke or standing at the barn door, have a good yarn till it is over? What is the mild air for, if not to enable us to wait half the day for the possible arrival of somebody or something whose presence or absence really makes very little difference to us? It is rare to see a man ploughing without two or three companions walking at his side, or a woman milking without a crony at her elbow to beguile her labour with talk. If Donald goes to give Murdo a hand with his job, it is rather a tongue or an ear he is offering. Murdo is delighted, but less by manual aid than by the charms of conversation. Put a man to work near the road, and all the world is a wayfarer looking for company. Put him to work on the loneliest and most inaccessible place you can think of, and the very stones will rise up and start talking.

If you want a crofter to cut your grass, or his wife to sell you a dozen of eggs, never blurt out your business at once; it is not good form. An hour's preamble, in which you touch upon at least a score of topics of general interest, is necessary to introduce the real purpose of your visit. Even then it should be treated in casual postscript form. Turning from the door, you should stand poised on one leg, and contemplating with dreamy eyes the farthest hill in sight, mutter, like the villain in a melodrama, an aside: 'Oh, by the way, Kenny, if you are passing my way, you might look in and cut a puckle of grass, would you?' If you see somebody vaguely flitting about the steading, or lurking in some dark doorway, do not go up to him and ask brusquely 'What do you want?' It is not good form. In any case he will only look blank and abstracted, as if completely uninterested in you and your belongings, and begin to talk of something else. At the end of an hour he will still be there. After some more irrelevant conversation he will ask you to sell him a fat wether or give him the loan of your horse. One day Herself was digging in her flower border. Out of the tail of her eye she saw Donald lurking near the stable, watching her. She dug on. He came a little nearer. Still she dug. Little by little he edged up to her, and began to talk about the weather. 'He wants something,' she thought. Exhausting the weather, he went on to speak of the rams newly arrived from Hawick. Then at last, in his gruff, rasping voice he said: 'Would himself get me a bottle of port wine for my mother. She is not keeping well . . . weak . . . could get some at the hotel, but poor stuff. Not strong enough.' Then, producing a crumpled pound note from his pocket, 'Pay for it now'. – 'Yes, I am sure he will. No, Donald, you needn't pay for it now.' Herself went on with her digging. Donald remained watching her, a vague detached expression on his face. There was a long silence, broken only by the chink of spade against stones. Then: 'I was wondering would I get one of those red cockerels.' – 'Yes, I am sure you can.' – 'I will be coming for my horse to-morrow.' – 'Well, if you are coming for your horse, perhaps you would take up a load of gravel for me, and you will get the cockerel home with you afterwards.' They part the best of friends. But all this takes time. And yet people say that on the West coast nothing is ever done.

From the study of climate, too serious and abstract for my idle unscientific pen, I turn with delight to the highly personal subject of the weather. Those who habitually live in a big town have no conception of the weather. For them it is shorn of all its terror and most of its power. It provides a convenient topic of conversation with strangers, and gives them something to grumble at when they have no better grievance. Independent of its caprices, they feel no livelier emotion than mild annoyance or faint pleasure. The August deluge that wrecks a cornfield, the June frost that blasts an orchard, will do no more than set the townsman hunting for an umbrella or an overcoat. The weather in town is like the tame constitutional king who, having exchanged the function of cutting off heads for that of opening bazaars, reigns but does not rule. For without the power of life and death, royal state is a mockery. How often must a modern sovereign, when opening a cats' home, look down upon the sea of silly faces, and envy the Queen of Hearts, who would simply have shouted 'Off with their heads!' and what's more, seen it done.

But with us the weather is no figure-head – it is supreme. The Gaelic language, abhorring the abstraction of the neuter gender, calls it 'she'. It is right; for in doing this it is only paying a just tribute to her power. To endow a thing with sex is at once to give it life and personality. 'Dang old stove!' I once heard a Cornish woman exclaim, when her oven would not heat; 'I hate he!' The stove, from being mere dead iron, became a living male abomination. Thus personified, the weather is powerful indeed. She moulds our lives, shapes our characters, and governs our work. We may grumble at her, we may use every known device to circumvent her, but her supremacy is acknowledged. Rebellion is useless, and in the end we realise that the highest wisdom is to adapt ourselves to her scheme of things and make the most of it. She is capricious, incalculable; but what are brains and ingenuity for, if not to cope with the wiles of such a ruler?

In all parts of Britain the weather is uncertain, so that agriculture, which is dependent on its caprices, can never be highly organised or industrialised. A farmer cannot arrange his work in advance as indoor workers can. In the midst of so much variation, rigid plans and theoretical schemes are of little value. A well-known general once said that in war things do not go at all according to plan; and the same holds good of the farmer's truce-less struggle against natural forces. Plans may be made, but they have to be revised, modified, or cancelled to suit the needs of the moment. The flockmaster in the folksong, who asserts at the end of every verse that 'tomorrow my sheep shall be shorn!' was an optimist, if not indeed guilty of impious presumption, since he did not add 'Weather permitting' or 'If we are spared'. The moods of the weather make the farmer an unblushing opportunist; and who can blame him? And so much does this opportunism become a fixed habit, that he sometimes extends it to matters less variable, with results unsatisfactory to himself and provoking to others. He waits to see which way the cat will jump; but the cat does not jump at all, but with a knowing wink curls up and goes to sleep. What then?

Of all kinds of British weather, the West Highland variety is the most changeable. Long settled periods, such as often occur in southern England, are rare. Quick, often violent, changes of wind, temperature, and sky are the rule. This is due partly to the influence of sea and mountains, and partly to latitude, for lying in a kind of no-man's-land between the high pressure of the conti-nent and the low pressure of the North Atlantic, the Highland districts are subject to a perpetual see-saw, producing frequent gales, sudden and heavy downpours, rapid frosts and thaws, and perpetual shifts of wind. Someday, when meteorologists have discovered the secrets of the Arctic, the laws which govern this apparent lawlessness will be disclosed, and the Highland weather be as easy to predict and understand as that of districts with a fixed wet and dry season. But that time is not yet; and from the farmer's point of view the Highland weather is simply incalculable. It is impossible to plan work ahead, except routine jobs like milking or feeding calves. On many farms where sheep are the main concern and arable of no account, it may be wasteful to employ any full-time man beyond the shepherds, so little steady work is there for him to do. To meet the rush – I speak relatively – of spring work or harvest, it is better to send round for casual help. To this happy-go-lucky way of doing things the crofting system is specially well adapted. It is true that your crofter neighbours have their own work to attend to, and that the sun which is ripening your crops is also ripening theirs. But with a little give and take this difficulty can be overcome; and if you can persuade yourself to be rather less procrastinating than they, there is a good chance of getting your work done before they have started theirs. And besides those who actually work their crofts, there are a number of people, male and female, who mostly exist beautifully, like the lilies of the field, but from time to time, when the mood takes them, deign to work as ghillies, extra shepherds, foresters, or household helps. These, if they can be prevailed upon to come to you every third day or so, may be of service. And if you are really canny, you will dangle in front of your summer visitors the Arcadian delights of haymaking, and point out to them that more healthy exercise can be got by forking hay than by brandishing golf clubs. With luck they may believe you.

We say to Angus, 'What about coming up the next good day, and cutting some grass?' He comes, and we saw some planks for his new cart. There is something pleasant about this neighbourly co-operation, where benefits are given and received without the passing of any money. Of course the system has its annoying side. Your neighbour is free as air; you have no hold over him, you cannot count on him. The good day comes, all is ready, but no Angus. His idea of a good day may not be the same as yours, or he may have found something better to do. There is no sense in enquiring, and still less in fretting yourself into a fever of annoyance. The only thing is to say, 'Well, that's that', and do the job yourself. A good many people who go abroad are willing to put up with any amount of discomfort and inconvenience with an amused smile, because 'it is all so delightfully foreign', whereas a far smaller dose of the same thing in Britain makes them writhe in anguish. Well, Skye is part of Britain, and a post office in Skye does much the same business as a post office in Essex. But Gaelic Skye is foreign, just as foreign as France. And unless we can find the Skyeman's peculiarities as tolerable and even delightful as the capers of Paris or Corsica, we shall never be happy in Skye. Do as the native does, even down to watching your wife hoe potatoes while you sit on a rock and smoke, for that is the way to contentment. However particular, punctual, and systematic you may be in England or in the lowlands, abandon it here, in the interests of peace and happiness. It is bad form, just as it would be bad form to wear a loin-cloth in Bond Street or a clean face among the Eskimos. Besides, to swim perpetually against the stream is exhausting and in the end may prove fatal. 'Fret not thyself because of the procrastinators' is the best text for a Highland sermon. For if man is unreliable, it is perhaps because he has to adapt himself to the weather that rules his life.

Wise men have learnt to view the vagaries of the weather with a philosophic detachment. We should probably be much happier if we treated the moods of our loved ones in the same spirit. It is raining now, but the sky is breaking in the west, and by afternoon I shall be able to work in the garden. Jimmy has been cross all day, but the cloud is lifting on his brow, and by evening he will be sufficiently amiable to listen to my proposal that the black heifer should be sold. It is freezing too hard to put the cows out just now; but never mind, I will give them hay in the byre. Matilda is in a rage, and has left the separator dirty; but no matter, I will clean it myself. Of course you will say that people have souls to save, and should know better than to behave in the fashion of that blind irrational power – the weather. But it has yet to be proved that the weather is as blind and irrational as she seems; and until we have the faculty of disentangling and comprehending the complex network of human desires and motives, the happiest attitude is philosophic calm. To an all-seeing eye, the principles that govern human conduct would be as plain as the fundamental laws of weather. If we understood these laws, we should watch an approaching storm with detachment, because, forewarned, we should have taken precautions against it. But we have not an all-seeing eye, and have taken no precautions. Damn! As I write, the cows are meditating a descent on the cabbages. What fool left that lower gate open? Probably myself. A most unphilosophic rage consumes me. Let us get on to some subject less full of humbug.

12

July: Prelude to Haymaking

JUNE ended, and with it went that spacious sense of leisure. The rains of May, followed by the exceptional heat of the past three weeks, had made the grass like a tropical forest; the sawmill field would soon be ready for cutting, and I hoped that an early start might enable us to make some hay before the usual mid-July break in the weather. This hope, as will be seen later, was doomed to disappointment; but at the time Peter and I were full of plans to get half our hay made by the first of August. Our chief anxiety was to secure a regular scytheman. Peter was handy with the scythe, but had not had enough practice to be able to keep working for any length of time. We had previously sounded Murdo Macgregor, who then seemed eager to come, but a rumour reached me that he had been offered regular work in the village. Good scythemen were scarce, and in great demand; if Murdo failed, I could think of no suitable substitute. It must be understood that cutting hay with a scythe is a skilled job, requiring strength, dexterity, and practice, and if the business is not to be hopelessly tedious and expensive, you must secure a man who can cut closely and steadily in wide sweeps all day, and knows how to keep his blade as sharp as a razor; for there are some who can cut well enough, but sharpen their scythes at such an angle that the edge is never good, and the whole blade worn away in a fortnight. The previous year, Angus had asked me for a job on the hay, and I had set him and Alec to mow a small paddock which now belonged to Rattray. The two boys took three days to cut one-and-a-half acres; so that at present-day wages I reckoned that my hay would cost about £8 a ton to produce, and stopped the proceedings!

The reader will probably be wondering why in the world we do not use a mower. The answer is simple. If machine-mowing could be combined with an artificial drying plant, all would be well, and you could mow acres a day without anxiety. But, in a climate where either it is raining today, or has rained yesterday or will rain tomorrow, the only correct method is to cut just so much as your haymaking staff can handle at a time. One good scytheman can keep two or three or even four haymakers busy, and can be made to drop his scythe and do something else at a moment's notice. Further, a mower is a machine, and the time lost in tinkering with it would cancel out any gain in the speed of actual cutting. The thing would work like this. A fine morning for the hay. The mower is taken out, the horses yoked, and after an hour or two of talk and chinking it starts. A round or two is cut; then comes a real or supposed breakdown, and talk and chinking is resumed, possibly till dinner-time. In the afternoon we start again, and all goes well until a shower is seen approaching, which perhaps will be no more than a passing scud, but may be and probably is the prelude to a heavy deluge. It is not safe to cut much more, so that horses are unyoked and the mower abandoned. And apart from all this, if we had a mower we must keep two horses, for it is not possible to borrow one regularly at this time of year. And to keep a second horse when you have hardly enough work even for one is obviously poor economy. No: in the Western Highlands, unless you can devise some artificial means of drying hay quickly, the scythe is the quickest and most economical method of cutting hay.

The most effective scytheman I ever saw was an elder brother of Murdo's. He had not the style and precision of some of the older men, but owing to his immense sweep and tireless strength, he could cut more grass in a day than anyone in the place. He was a man of great bodily strength, who could carry on his shoulder logs weighing three hundredweight and more without turning a hair. The Laird had known him since childhood, when he used to pass through Strathascaig on his way to school, an incredibly ragged and dirty little urchin, whose way of descending a steep brae was to lie flat and roll down. Later he joined the regular army, but was dismissed for fighting, which the Laird thought a little odd, because fighting is said to be the army's chief purpose. After this episode he came home, living officially on the proceeds of casual labour, and unofficially on what he made by various secret activities. Fighting was still his chief delight: not long ago he was summoned for assaulting an Indian pedlar, borrowed £2 from the Laird, who is a JP, to attend the court, and though guilty on his own admission, succeeded in getting himself acquitted, chiefly because the pedlar had very little English and his interpreter garbled the evidence. He had a smart carriage and a winning smile, and his strength and resource in devilment would have made him an excellent pirate or brigand, had these attractive and profitable callings still been generally available. As it was, he had to content himself with spells of work on the roads or wood-cutting for the Laird; and in the intervals he would disappear for days or even weeks to spend the proceeds on riotous living. He had a turn for dress, and last summer arrived to cut the hay in so impeccable a pair of grey flannel bags, complete with sports shirt and Fair Isle pullover, that I mistook the radiant vision for a strayed tourist, and only when he picked up his scythe did I recognise the most elegant of the Macgregors. He did not work long at Achnabo, for I refused to employ him without an insurance card; and as he despised such quibbling trifles, he quickly disappeared and sent Murdo in his place. This was the beginning of Murdo's connection with the farm; he had worked with Alec at the hay throughout the season of 1933, and as he had proved a good and reliable scytheman, the prospect of not having him again this year was rather annoying.

I could get no certain information, and one wet afternoon I walked over to see him. In no-man's-land I found him with his young brother, catching their pony. He said that he had applied for this job in the village, but had heard nothing definite, and would come over as soon as he knew for certain. The job was a permanent one, and if reasonably well paid would be better than casual work: Murdo was the only Macgregor who had any desire to be in regular employment. A few days later Dan, the young brother, turned up with a letter in a clean and well-written envelope, carefully marked in the left-hand corner with the words 'By Hand', in which Murdo, who was a good scholar, stated that he had been offered 11d an hour with his food, but he would rather come to me, if I could see my way to paying him 10d, or a penny an hour more than last year. To this I agreed: it was rather a high wage, but the standard had been set in the district by the Forestry Commission, and it was difficult to get labour worth having for less. For various reasons he could not start work till 23 July, and sent Dan in his place; but this lad had never scythed anything but bracken, and there was so much slashing and chewing and tinkering with the set of the blade that I got rid of him, and Peter worked alone.

The first hay was cut on 10 July, when Peter rose early and mowed a long strip before breakfast. It was a very hot day, brilliantly fine, with no prospect of a change. Little did we think, as the first swathe fell, that we should take four months to make less hay than we had the previous year, and that after 16 weeks of almost incessant rain we should carry the last load on 9 November! This cut was, however, more by way of practice, for we had not yet finished our arrangements for drying the hay indoors. For the benefit of those who have not seen hay made on the West coast, I will say a word or two about the best method of securing it. In spells of fine weather, which are of course comparatively rare, hay is raked and turned on the ground in the usual way. But if there is any danger of rain, it is put on wire fences, of which more later, and then when half dry is carted under cover, where the drying process is completed on specially made racks. For this purpose the sawmill, which measured about 70 feet by 18 feet, was excellently suited. Before I came to the farm, it was mostly open at the sides, in the fashion of a Dutch barn. But in windy weather the rain would drive right through it, and the problem was how to keep out the rain while admitting enough air to dry the hay. We solved it by cutting a large number of straight young pines, with trunks 1½ to 2 inches in diameter. These, when set on end with about an inch of space between each, made an excellent ventilated screen, through which the wind would whistle merrily, while little or no rain could penetrate. We cleared the building of all trash, and in the middle set up a drying rack, about 20 feet long by 12 feet wide, standing 9 feet high and the back end sloping to the floor at an angle of about 45°. It was made of flat spreading spruce branches supported by light yet strong poles, which were hinged with wire to the beam at the back, so that the rack could be hauled up under the rafters when no longer required, and let down again in preparation for the next haymaking. It was this letting down of the rack that kept us busy on the morning of the 10th. Peter climbed into the rafters and cast loose the lashings, having first attached fresh ropes by which the rack could be steadied and finally lowered. These ropes were worked by Mr Gordon and myself, and with various jams and hitches the light but ungainly structure was at last successfully lowered and made ready to receive its load of hay. The two ends of the sawmill were left open for packing the hay when dry, while the space in front of the rack was cleared for forking it on and off and storing loads of half-dry grass as it was carted in from the field.

We had made a second rack in the barn below the house. But in the previous autumn Dick and two stirks had climbed in through the window and in their efforts to get at some potatoes stored beneath had broken the rack and trampled it to pieces. The posts were still sound, but the spruce branches completely broken, and must be replaced. Actually this rack was not mended till the 21st, for until Murdo's arrival the sawmill was enough for our needs. The 21st was pouring wet, and as work on the hay outside was quite impossible, we took axes and saws to the wood by the road, where spruces were growing in accessible positions. Peter climbed the trees and sawed off branches all round, while I dragged them out to the road. Here we piled them into the form of a sledge, and towed them home with ropes through the pelting down pour. We had to make three journeys, and it would have been quicker as well as easier to have caught the horse; but he was in the lower pasture, and with characteristic West Coast laziness we made more work in seeking to avoid it. We were toiling up the brae to the farm gate when a smart car bore down upon us, with its owner in all his glorious clothes, going off somewhere to amuse himself. We took an evil pleasure in making him sit with his engine running while we slowly dragged our uncouth load up the brae and into the steading.

To go back to the 10th. Hot as the morning had been, the afternoon was even hotter, and by evil fortune we had to get our six stirks to the station. They were entered for the sale on the 11th, and were to be trucked by the 6.20 a.m. train, but in order to avoid any trouble and delay in the early morning we had arranged to drive them down on the afternoon of the 10th, and leave them overnight in the school playing-field beside the station, from which they could be quickly gathered and put into the truck on the morning itself. They were the first beasts to be sold off the farm, and I felt sorry to lose them. There was Theodore, a Guernsey Shorthorn cross, and Peter, a cross-bred Galloway, both two-year-old bullocks bought from the Laird, with three yearling cross bullocks and one yearling heifer reared on the place, called Bede, Hannibal, Brutus, and Bella. To these we had added a blue-grey polled cow with calf at foot, the detestable Augusta, who had been responsible for every gate-crashing and fence-breaking incident that had happened on the farm. Apart from this vice, she had poor-quality milk and did not make a good job of her calves. Though not a young cow, she was a good beef type and in excellent condition, and should have made a good price.

At half past three Angus came up to help Peter with the cattle. The heat was overpowering, and Angus, like most dwellers in temperate climates, was completely prostrated. Things were not helped by his devotion to thick underclothes. Last summer, when he came up to mow Rattray's paddock, we were afflicted by another but much briefer heat wave. As the morning wore on and the power of the sun increased, Angus wilted visibly, and I discovered that he was wearing a woollen jersey, a thick khaki shirt, and under that an even thicker and badly felted woollen vest! He had discarded the jersey, but though the temperature must have been over eighty in the shade, he still clung to the shirt and vest. I suggested that at dinner-time he should abandon either the shirt or the vest. He took my advice, and in the afternoon was considerably revived. On this last occasion he was rather less heavily attired, but even so must have carried double the weight of Peter's cotton shorts and shirt.

The beasts for the sale had been put in one field together, so that they might get over all preliminary fighting and be well settled down before trucking. They were easily gathered and driven down the road, but they were feeling the heat and tormented by swarms of clegs and flies of every description, so that it was necessary to rest them halfway. The road to the station was unfenced, and there was a certain amount of breaking away and running about, but much less than was anticipated, and the journey passed without incident.

Next morning we rose at half past four and, accompanied by Mr Gordon, started on our walk to the station, thus leaving plenty of time to take the cattle out of the playing-field and into the truck, which would be waiting at the siding. When the sun was high the day would be as hot if not hotter than yesterday; but at this early hour there was some freshness in the air, and walking was not unpleasant. At the point where the Achnabo land is divided from the crofters' common grazing there was a gate across the road, which we tried to keep shut, partly to prevent neighbours' cows from straying onto our ground and partly to annoy motorists, who would thus be forced to stop, open the gate, move on the car, stop again, close the gate, and then proceed. The gate was closed, and on the farther side I could see a group of cattle, presumably the crofters' cows, waiting for a chance to get in. I was just going to set the dogs at them, when I recognised Augusta's grey complacent form, and knew that the beasts were our own. How long they had been there I do not know, but probably they had escaped from the playing field the previous evening, and had spent the night at the familiar boundaries of home. It was more than fortunate that the gate was closed, otherwise they would have gone straight back to their old pastures, which were out of sight of the farm, and we should have reached the station without discovering our loss. How they got out of that playing field remains a mystery. The fence was new, strong, nearly six feet high and protected with barbed wire. Even Augusta would have been puzzled to find an exit. I cannot help thinking that somebody must have let them out, though why, I do not know.

So all the effort of the previous afternoon had been wasted. Fortunately we had plenty of time, and in the cool of the early morning there was no need to rest halfway. We reached the station forty minutes before the train was due, and by the judicious use of hurdles got them all into the truck in five minutes or less.

We were home in time for the usual routine of milking and calf-feeding, and after breakfast resumed haymaking in earnest. Peter had not been long at work before the Laird's shepherd appeared with one of the local crofters, ostensibly to gather any Strathascaig sheep that might be on our land. There were no sheep, but they stopped to watch Peter at work. The shepherd as usual was content to sit and advise, but his companion, who was a skilful scytheman, opened a new cut for Peter in return for a dairy pailful of water. With the increasing sunshine the clegs had come out in force, and by eleven o'clock they were so intolerable that Peter was forced to exchange his shorts for trousers. I put on gloves and three pairs of stockings, while Mr Gordon, who was helping to turn the swathes, had in addition to gloves and trousers a large Australian hat, and his neck swathed with handkerchiefs steeped in citronella. Even this availed little. Clegs are often troublesome in July, but I have never known them as bad as they were this summer. Their slow, secret method of attack, their persistence, their painful bite make them an unqualified nuisance. It is true that they are easily killed, but as one is rarely conscious of their presence until the bite is actually felt, the satisfaction of killing is greatly diminished: to prevent an injury is better than avenging it.

When the clegs went in the midges came out. There is no doubt that insects have been the cause of much atheism; and no philosopher or theologian has ever been able to vindicate their right to live in a way that will satisfy the ordinary man. Perhaps in Nature's wasteful fashion the myriad noxious species are there to make possible the existence of a few that have use or beauty. But bees have stings and butterflies have grubs, so let us be bold and damn the lot, at least until some wider knowledge and more divine forbearance has made us probe beyond the bites and buzzings to some inherent harmony of purpose.

In a polite book, a paragraph on midges would be a set of blanks, dashes, and asterisks. Few writers on the Highlands fail to mention midges, and I can explain the temperance of their language only by supposing that the publishers have bowdlerised the text. This is not a polite book, but to save ink I will leave out the swears; the reader must take them as being said in a perpetual crescendo of exasperation for ever and ever, world without end. It is well known that in the north, where the summer is short and surprisingly hot, the god of flies enjoys a brief span of fierce and feverish activity. As it says in the Bible, 'The devil is come among you, having great wrath, knowing that his time is short'. I take it on authority that Highland midges are not as bad as those of Lapland, Siberia, or Labrador, but they are bad enough; indeed it is sometimes difficult to imagine them worse. There are still evenings in June when the whole world is like a vision of paradise, but it is impossible to stay five minutes out of doors, except well out to sea. Midges cannot stand strong sunshine, and on cloudy days, especially of the muggy windless type so common in July, they are supreme. So maddening are their attacks that we are sometimes compelled to stop all outdoor work. (This is their only saving grace.) The most powerful tobacco fumes, the most acrid camp-fire smoke, the smelliest lotion concocted by chemists, are all equally useless. People often ask what midges feed on when there is no one about. This question is purely academic; it has as little practical importance as to ask if anything exists apart from our perception of it. If we do not perceive midges, for heaven's sake leave them alone. Apart from human blood, their diet can interest no one but the professional biologist, who, embalmed in laboratory stinks, is safe from the pest's innumerable arrows. There is a honeysuckle at the Laird's gate, laden with sweet sticky blossoms; and on still days these flowers are black with midges. Shake a branch, and a cloud of them will rise up with a thin and maddening buzz, but so intent are they upon their sugary feast that they will not bother to bite. To wreathe oneself in honeysuckle might be a safeguard, or to make a panoply of butterwort and honeydew. Perhaps. Or just to sit indoors with a book, and let the hay and the garden take care of themselves. For ten minutes outside means two hours of feverish scratching, and a catalogue of damns unrivalled by the bloodiest seaman. Is it worthwhile? I hardly think it.

The superfluity of midges is luckily balanced by a scarcity of wasps. I cannot remember the time when I did not detest wasps. Too careful, or if you like too cowardly, to get stung, I am yet more afraid of wasps than of adders. I loathe their gaudy black and yellow, their minute waists, hollow restless bodies, twiddling antennae, exasperating buzz. As a small child I vividly recollect the agony of being forced to sit still and listen to one of my uncle's interminable sermons, while the horror, droning like a hostile aeroplane, hovered over the rectory pew. At the age of six I met a man who could kill wasps on the wing with a flick of his fingers. Deeply impressed, I made him promise to marry me. That promise was never fulfilled, but as he died a bachelor, I fancy that I was the only woman who ever discovered his real capacities. I have now acquired sufficient self-control to sit quietly at a banquet shared with wasps, and even to eat, though without much relish, and to talk, but with little intelligence. Had I all the wit and wisdom in the world, I could not discuss a serious subject with a wasp crawling on my plate. And in my worst nightmare, I bite into a melting, succulent Victoria plum, and find it full of black and yellow wriggling horrors.

On 12 July the barometer fell rapidly; it was obvious that the weather was going to break. The air was hot and oppressive, the sky veiled with murky clouds, and later in the morning thunder was heard far away to the eastward. (It is characteristic of our district that the winter thunder squall comes from the west, while the orthodox summer storm, when we get one at all, works up from the east.) The swathes of hay, which were hardly dry enough to take in, were hastily forked into coils which, if the rain held off, could be spread again next day. They never were spread: it rained incessantly for the rest of the month, and the sodden coils were finally carted to the fence and dried there. With the exception of one or two isolated attempts in August, no more hay was made on the ground that year. The coming of the rain was not an unmixed evil, for it put an end to the clegs, and filled up our cistern, which had fallen so low that the water was not pleasant to drink, and was probably the cause of an attack of fever and sickness that put Peter out of action for some days. He had drunk nearly a pailful of unboiled water while scything in the heat. The same week we lost the bull calf Wilson, who died of some stomach trouble eight days after birth.

On the 13th the cheque came from the salesman. I had not been expecting much, but the result was disappointing, especially as the previous week's prices had been quite good. The six stirks made only £41 between them, and Augusta and her calf £13. The two-year-olds, which had been bought as yearlings in the spring of 1933 at £10 each, fetched only £8 and £9 apiece; thus with the cost of wintering, they made a loss of £6. The yearlings were home-born, and so realised a small profit. On the whole transaction I was about square, for the cow, which had cost me £16 in 1933, had produced two calves on one wintering. The beasts were in prime condition, but the prolonged drought on the eastern side of the country had spoilt the turnips and caused a deficiency of grass for mowing, so that farmers were not anxious to buy many stores for fattening. The terrible heat on the actual day of the sale would not have stimulated the bidding nor improved the appearance of the animals. Had we kept them back another week or two, we should have reaped the benefit of a short-lived rise in prices that followed the publication of the Beef Subsidy Scheme.

That week we had no butter, for the thunder on the 12th turned the cream sour before it could be churned. There are people who say that thunder has no effect on milk or cream, but in actual experience I have always found that the process of souring is more rapid when there is thunder about, even when the thunder is not associated with very high temperatures. And the Achnabo milk was a fairly representative specimen, being produced in reasonable cleanliness but not under Grade A conditions. We have never sold milk officially, and I suppose that, if we do, it will have to be in conformity with all the latest clean milk shibboleths. The question is far too thorny and complicated to be discussed here, and all I shall do is to make one or two remarks which may have some bearing on the subject. First of all, why can we not be honest about it? The object of producing clean milk is not to save the consumer from illness, but to improve the keeping qualities of milk and so make more money for the producer and retailer. The modern urban housewife has a superstitious reverence for science, and the talk of low bacterial counts is music to her ear. She pays to have sterile milk delivered in sealed bottles to her door, and then as likely as not leaves the milk standing in a not altogether sterile jug in a larder exposed to the smuts of London. The only safe thing would be to pump it straight from the cow's udder into the consumer's stomach. But is the consumer worth it? Apart from these fantasies, if some method could be devised by which surplus summer milk was stored under vacuum and released in winter for sale at winter prices, there would be much profit in it, and a society which prefers New Zealand butter, Australian eggs, and chilled Argentine beef to the fresh products of British farms would find no hardship in drinking June milk in December provided that it tasted fresh.

At one time, and not very long ago, fresh farmhouse eggs and butter were esteemed in towns, and bought at a higher price. The very fact that they were not produced under standardised factory conditions was deemed a merit. Now such goods find favour only with a few die-hards and sentimentalists, while the majority of consumers, guided by the directors of large creameries and the more academic of the medical profession, share the views of a small boy from London who went to the country and refused to drink his milk because it came from a nasty dirty cow instead of out of a nice clean bottle.

To take all our food completely sterilised would be well enough in a wholly sterile world, where immunity from infection had no meaning, since *ex hypothesi* there would be no bacteria to infect. But as things are, when even in the cleanest place they swarm by the million, the consumption of sterilised food must lower our powers of resistance. With such methods we are only getting further and further from Nature, and by the elimination of natural selection, producing a race that in the end will only be fit to live in glass bottles on a laboratory shelf. Even twenty years ago, the number of men rejected as unfit for military service was beginning to alarm those who cared for the physical welfare of the nation. At the present day, millions are being spent on keeping alive people who should never have survived infancy, and dragging them through a life of debility to breed their like and begin the whole dreary cycle over again. An incorrigible cynic might suggest that our doctors would profit by a world in which half the population were patients; but without going so far, one might speculate whether a fair proportion of infant welfare work was not, on the long view of national health, a serious mistake.

No farmer in his senses would build up and maintain his stock as we do our race; and yet the welfare of the human community is worth more than a herd of cattle.

From these reflections I turn with relief to the Belties. They were all in prime condition, over two years old and ready for service. We put the bull to run with them on 30 June, and there he remained for six weeks, so that all were expected to calve sometime in April or the first half of May. At this time there would be a bite of grass and short nights, so that they need not be taken indoors at all. George was in clover. Sleek, shiny and utterly contented, he would lie chewing his cud, surrounded by his shaggy and devoted harem. In a few months their clumsy heiferish gambols would cease, and as the calves grew heavy within them, they would settle down to the placid, monotonous round of grazing, calving, suckling, and mating, at peace with themselves and a profit to their owners. As they moved about on their sturdy little legs, sweeping the ground with their tails and twitching the flies from their softly feathered ears, I fell in love with them completely. Their gentle eyes peered forth from a fringe of hair as long and tufted as a Skye terrier's, and their jaws revolved ceaselessly as they chewed the sweet grass of the summer pastures. I should not mind working day and night at the corn, to keep those jaws revolving ceaselessly all winter. Six months on, and they would be standing at the fence of the wood with a powder of snow on their backs, rending the yellow sheaves and chewing for dear life, with wisps of straw stuck jauntily out of the corner of their mouths. And in April, in the long solemn twilights, austerely cold but full of the resistless onward march of spring, they would be casually dropping their calves in the thickets, licking the weak wet creatures with their harsh tongues, till the blood began to circulate freely, and they rose on long unsteady legs to suck the hairy teats. In a few hours they would be racing about, full of warm food and the zest of life, under the watchful placid eyes of their mothers, and so the cycle would begin afresh.

About this time we saw an advertisement of a tripod device for drying hay and corn, and after much study of the folders sent at our request, we ordered some larch spars from the local sawmill, with two rolls of fencing wire, and made a set ourselves. The hay tripods were very useful, but as it turned out, the harvest weather was too wild to get much benefit from those intended for corn. The idea of the tripod is to support a coil of hay or corn in such a way as to leave an air-space in the middle, with ventilating tunnels on three sides, through which the wind can pass and keep the stuff sound and cool at the heart. The tripod is a modification of the ventilating boss round which permanent stacks in the north are usually built; the only new thing about it is its use for small temporary coils in the field. The makers claim that hay and corn can be built into 'huts' round tripods the day after cutting, and dry out completely in the hut. In practice, we found that semi-green grass would stand for several days, even for two or three weeks, without heating or deterioration, and if there were plenty of wind, would become appreciably drier, or at all events easier to dry; but it was always necessary to shake it out and spread it on the ground before carting it indoors. In the case of corn, the stiff straw prevents the material from settling very closely, so that there are always interstices left through which some air can pass; and for this reason the ventilated coil is not so greatly superior to the ordinary solid hand-rick of similar size and shape.

13

Interlude on Cattle and the Art of Driving Them

AS a small girl I used to attend Matins at a suburban church full
of pious old ladies and reluctant children. In Scotland the service
would have been considered too short; but to us youngsters it
seemed interminably long, and afterwards we would drearily ask
one another which we hated more – the Litany or the Athanasian
Creed. In this desert of dullness there was one green tract – the
Psalms. And in the Psalms there was one verse that delighted me
above all others, where God is made to say, 'Mine are the cattle
upon a thousand hills'. There was, and is still, something in those
words that fires the imagination; 'something mighty', as an old
Cornish farmer once put it, 'something tremenjous!' A vast sweep
of rolling downs, dotted with cattle innumerable, black, white,
brown, speckled, male and female, horned and polled, the immense
wealth of the patriarchs gathered in living form for all to see.
When in later years I beheld from the Gulf of Suez the barren hills
of Sinai baking in the sun, these pastoral riches seemed even more
striking in a land of thirst and desert, where, at least to my rain-
blurred eyes, there seemed so little food to keep the beasts alive, so
little shelter to protect them from noonday heat or the chills of
night. To the simple mind, wealth in visible or tangible form is far
more impressive: at bottom we are more stirred by the jingle of
sovereigns and the sparkle of diamonds than by the most opulent
crackling of banknotes or suave smoothness of dividend warrants.
What clearer token of wealth could we demand than a huge
moving forest of horns and switching tails, alive and vocal, and all
our own? And what possession could be more dignified? 'Mine

are the cattle upon a thousand hills.' No modern Jew would dare to make his God exclaim, 'Mine is the capital that finances a thousand commercial enterprises.'

In the old days, cattle were the standard of real wealth, as gold is today. With cattle, men bought and sold their most precious possessions; and in these primitive communities, as in the west of Ireland today, a girl was marriageable in proportion to the head of cattle she could bring with her as a dowry. Now, at least outwardly, things have changed. A millionaire's income, even if derived from Bovril or condensed milk, is counted not in cows but in dollars. A woman may own a pedigree herd of Jerseys, and never a suitor cross her threshold. The sealed bottles that arrive daily from the Grade A dairy, the leather attaché case we take with us to the office, the rissole we eat for lunch at Lyons, are barely recognised as animal products. Cynics will tell us that they have been through so many processes that their origin is quite lost sight of, if indeed they are not entirely synthetic; but never mind that, the argument will work all the same. The useful beast is kept well in the background of our lives. Perhaps it is all for the best. Townsmen, do not enquire into the sources of your civilised products: that way madness lies. Press the button and ask no questions. But however much in the background, except when they obstruct motorists, cattle are still indispensable, as they were in ancient days. Few modern developments are more striking than the chilled-meat industry of Argentina. And meat to be chilled must once have been warm. In our own country, much time is spent by politicians and journalists in talking and writing about the decline or the revival of agriculture. Armies of government officials are employed in regulating the sale of milk and the movement of livestock, and compiling detailed statistics of the farming industry. Every schoolboy will soon know how many millions of bacteria he drinks daily in his morning milk.

All this interest in agriculture, though a little academic, for very few of these people could milk a cow or make a stook of corn, shows that cattle are still important, even in cities. But you must go to the country, where people actually keep them, to discover how their habits and necessities rule our daily lives. Railway time-tables, hours of church services, arrangement of meals, are all dominated by the needs of cattle. 'I can't meet you till after nine, because of the separator. Dearly as I love you, it is impossible.' 'We must finish this rubber another time; the cow has begun to calve.' 'I must not discuss the Gold Standard with you any longer; the calves are in the corn.' 'Your remarks about Predestination are most interesting, but excuse me, I must go. My neighbour is bringing his cow for the bull.' It is largely the demands of cattle that prevent the agricultural labourer, who shares this disability with the mother and the housewife, from getting his Sunday off. So that it might not be a waste of time and money if the Trades Union Congress, instead of dabbling in high finance, subsidised livestock breeders to evolve a six-day cow that would skip Sunday and produce a double quantity on Monday.

Most people have heard how, in days gone by, Highland girls used to go with their cows to the shielings on the hills, where they spent the summer making butter and cheese. This Arcadian life, together with the people who cared to lead it, has gone beyond recall: the shielings have fallen into ruin, and the milk-maids are filling the mills and cinemas of the south. Within the limits of the glens, an abundance of good pasture, rich in natural clover, is still be to found; but the small population cannot even supply its own needs, since in certain districts, notably in Lewis, milk is imported from the east coast.[1]* Dairying can never flourish in the West, for there is neither an outlet for surplus summer milk, nor any supply of suitable labour. In Scotland milking is a woman's job; but very few girls remain in the country, and those who do so think it degrading to handle a cow. Why milking should be a less dignified occupation than cleaning a stove or washing other people's dirty plates or linen I have never been able to fathom; but I suppose there is as much to be said for it as for the

[1] Written in 1932.

equally widespread notion that no water is pure unless it comes
out of a tap.

On the other hand, the old industry of cattle-raising, which
now takes the form of rearing calves of a beef breed to be sold as
soon as weaned, has in all probability a long and profitable future.
It is suited not only to the country itself, but to modern conditions of labour, transport, and marketing; and as long as sheep
farming continues to be the chief industry of the Highlands, farmers will find it worthwhile to keep a certain number of rough
cattle to eat down the coarser grass. For many years before the
present slump the price of weaned calves had been steady and
good; and there is little fear that the demand for them will grow
permanently less. The annual sale of one or two calves provides
the crofter with the chief part of his income. Breeds, however, are
changing. The number of Highland cattle is steadily declining,
and it is not unlikely that in a few years' time these shaggy and
picturesque creatures, kept only to frighten trespassers in English
parks, will become as rare as the thick mists in which painters love
to envelop them. Not merely are the Highlanders slow-maturing,
but they are not very hardy; their curly coats do not run off the
rain, and they require more food in winter than certain other
types. In crofting townships they are being rapidly superseded by
Shorthorn crosses, while large farmers, who run a certain number
of cattle in combination with sheep, prefer the hardy and profitable Galloways.

At Strathascaig the Laird had two groups of cattle – the
Guernseys, imported from Cornwall to supply the house with
butter and milk, and the Galloways for breeding calves. The milk
cows were so few that we had little of the standardised routine of
the big dairy farm. We looked after them ourselves – milked them,
herded them, fed them. They helped to make our daily life, and as
often as not our daily cross as well, for few things are more irritating than an obstinate cow. That is why even well-conducted
farmyards ring with strange oaths, and humane farmers break
stout sticks into splinters upon their stubborn backs. Sailors are

thought to be hard swearers; but if all the ropes, spars, and blocks on a four-masted barque got jammed at once, it would not be more exasperating than when a dozen cows go off in different directions, each bent on eating some strictly forbidden fruit. Yet we are fond of them. They are not, of course, companionable, as dogs are, but you may love a cow without wondering whether you are developing a maternal complex. They are too large, and often too dirty, to be taken to your bosom; you cannot greatly indulge your emotions, except perhaps the angry passions, at their expense, and even if you make money off them, it is done with much toil and anxiety. We do not, I say, love them as we love our dogs; but we think of them with compassion when a premature storm overtakes them while still in the field by night; and when in early spring we see a green bite of grass appearing under some sheltering dyke, we wish they were there to enjoy it.

And the very quality that most annoys – their slowness – can give us, if only we would banish hurry, a sense of spaciousness and peace. The deliberate movement of the herd has a patriarchal dignity, as if we had left the crazy whirling of speedboats and racing cars, and returned to the solemn ox-waggons in which the Boer pioneers, with their wives, Bibles, and children innumerable, made their way across the boundless veldt. Domestic animals, too, are slaves of habit, and in the ebb and flow of human emotion this rigid conservatism may bring us comfort. It was once said that there were only two fixed points in a changing world – death and the servants' dinner. Yes, but to this I would add, or rather substitute, for only a snob would think the servants' dinner inevitable and ours merely contingent, as if hunger were a matter of class – to this I would add, milking-time. Your heart may be broken and your limbs failing you, but there the cow stands waiting and you must sit down and begin. Draw the milk from her teats, and she will turn her head and gaze at you tenderly, as if you were no longer merely owner, but a human substitute for the calf she has lost. And you will rise from the stool refreshed.

There were five Guernsey cows, three of them originally

imported from Cornwall, besides young beasts and a home-bred bull. They were five days on the railway, these pioneers, coming from a windswept upland farm on the north coast of the Penwith peninsula, from which it was thought they would easily acclimatise to the rather similar conditions of Wester Ross. Of these the leader was Jennifer, large, placid, round-bellied, of a pleasing pale fawn colour. She was the nearest approach to the economic cow dreamed of in dairy research institutes, that perfect contrivance for turning grass and hay into butter and milk systematically and without waste. When entering a new field, she never lost time in investigating its contents, or in gazing inquisitively at passers-by on the road. Not exploration nor pleasure was her bent, but filling her belly. From dawn to dusk she would graze steadily, with an interval at noon to chew her cud. Psychologically this is very dull. But from the farmer's point of view it is most desirable.

Quite different was Pendeen, whom later I brought to Achnabo. She was a brown and white Guernsey, slightly built, with a dark muzzle and unusually long eyelashes. Quiet enough to milk and handle, she had other interests in life besides converting the nearest available green stuff into milk. A man passing along the fence, the weeds that the gardener had thrown over the wall, sheaves of corn waiting to be stacked, leaves of sweetbrier in the hedge, an empty pail, a pair of stockings on the line, the piquancy of nettles, intriguing docks and thistles – all these things held more interest for her than the plain grass at her feet. She was a born investigator, explorer, pioneer. Above all, the tempting, mysterious thing on the other side of the fence attracted her. Were she human, this quality would be called the love of adventure, romance, idealism, or the reforming spirit. But she was only a cow; and we called it a nuisance. So it was.

An old woman I knew had for her garden gate the head of an iron bedstead. Inside the gate was a particularly succulent bite of grass. One day a questing cow put her head through the interstices of the ironwork, secured the grass, but was unable to get free again. One frightened toss was enough to wrench the bedhead from its flimsy hold, and the cow galloped off with the thing round its neck, ornamental as an Elizabethan ruff; and there it remained until someone pulled it off like a horse-collar. Another cow, in search of some other El Dorado, walked into the open door of a house, and being unable to turn in the narrow passage, must have climbed the stair, for she was next seen looking out of an upper window. And a third explorer, roaming the streets of Peebles, broke the window of a jeweller's shop and ate a bead necklace: nor was it easy to keep her from revisiting this choice and interesting pasture. The man exclusively set on domestic comfort will not, if he is wise, marry a wife who is interested in outside things, for even the most versatile of women cannot be all things to all men. So also the prudent farmer will not keep a wandering Jew of a cow, one who, like Pendeen, walks the world not in search of pasture, but for the mere pleasure of seeing all that is therein. Not long ago, in a newspaper correspondence about the origin of that fashionable verb 'to hike', it was stated that North Country farmers used to speak of 'hakers', or cows who would never settle to steady grazing. Beware, then, of hakers. They may amuse, but more often they annoy, and will end by emptying your milk-pails and last of all your pockets.

Of all the various jobs that fall to the cow-keeper's lot, by far the worst is driving. People who meet a small child in charge of a big herd of cows on a fenced road, think 'How simple! Any fool can do it.' *Can he?* It all depends on the cows and on the country. In flat fenced places, provided that the cattle have the herd instinct, so that where one goes the others follow, all is well. But in rough unenclosed country, where cows, having the Celtic individualism, must be driven as units, the driver should combine the speed of a hare, the resourcefulness of a sailor, and the patience of Job. He must also possess a pair of gumboots, lungs of leather, the vocabulary of a Yankee skipper, and a large selection of tough hazel sticks.

To convince the sceptic, I will give a few instances of problems of driving to be met with at Strathascaig, where the grazing was far less conveniently situated than at Achnabo.

First, the open hill. This may sound more formidable than it actually is, for cows, unless they are looking for calves or for the bull, do not roam far from their habitual feeding places. But the hill is rough, and when you go out to seek them, they may be lying in a hollow, screened by rocks, bushes, or high bracken, and completely hidden from view. In summer, reluctant to leave the pasture, they will spread out fan-wise over a wide front; you urge on the extreme right only to find that the extreme left has come to a halt. You rush across to the left: while they advance, the right halts. Thus, if unprovided with dog or other helper, the wretched driver races to and fro like a collie at a sheep gathering, and at last wins home, hot and exhausted, with a cracked voice and several broken sticks.

Then the moss. This, though fenced, is a wild enough place, specially in winter. It is a boggy plateau seamed with deep peat cuttings full of water and overhung with bush heather. Between the cuttings are stretches of soaking moss and lumps of coarse tussock grass. In summer it is beautiful with bog myrtle and waving plumes of cotton grass, and at all seasons has a great and individual charm; but from the practical point of view, the going is atrocious. The plateau, where it drops steeply to the farm road, is fringed with oak trees, whose ubiquitous roots make the slope all the more treacherous. The accumulation of surface water drains down through deep ravines, whose sheer gravelly sides are edged with young birches and hazels. A bad place for driving cows or leading horses.

Third, the flat. Like most of the pasture land at Strathascaig, it is separated from the steading by the river, which must be forded by the cattle. First thing on summer mornings I would draw back my curtains and scan the flat for the cows. If in sight at all, they were quickly picked up, for their light colouring was in contrast with the dark green of the bracken thickets where they loved to spend the night. But unfortunately the whole of the flat was not visible from any one place, and often enough I went out for the cows without knowing where to find them. The flat is an irregular space in the glen bottom, enclosed between the river and the road, and running to an extremity on the shore of the loch. It is intersected by one large burn and several small ones. Its higher levels, where it slopes up to the road, yield plenty of good grazing; but the characteristic flat part lies very wet, and is covered with coarse sedgy grass, rushes, and moss. In summer it is bright with the yellow flowers of the seilisdeir or wild iris, from whose roots a grey dye can be extracted; and the rare and beautiful grass of Parnassus, with its delicate green-veined white blossoms, is found also. Near the river are muddy ditches and pools abounding in frogs, whose vibrating croak sounds deceptively like an approaching motorcycle. Nearer the sea are strange shallow pools of brackish water, which in dry weather and at neap tides disappear, leaving a bottom of hard mud scored with a network of cracks. Nearby is the debris of the last spring tide, twigs, leaves, and an occasional tin, and round about, in May and June, the ground is covered with sea-pinks.

To the idle rambler the flat is full of wonders. But as a place in which to find cows in a hurry it is abominable. Some misty morning in September, when the season was advanced enough for us to catch the last of the dawn magic at seven o'clock, I have discovered with a shock of pleasure those green islands in a sea of dew, marking the spot where the cows had lain at night, and went along dreaming of Tess and Angel Clare in the meadows of Talbothays: until with ugly jerk into the present I thought, 'That is where they *were*; but the point is, where are the wretched brutes now?' Tracks in the dew, so beautifully described by Hardy, led to the river-bank, and there stopped. They had crossed, and crossed at the wrong place, too. There they were, embedded in a whin thicket, their horns and white-shielded foreheads half-seen through a bristling curtain of green spines. I went after them. Blows and curses resounded, breaking the magic of the dawn. Whack, whack. They turned, and began to amble solemnly homeward. I followed, hot, scratched, and weary. Then suddenly without warning, madness seized them. They all went off at a

clumsy grotesque gallop, heads down, their heavy udders ridicu-
lously swinging. In a few seconds they were far away. Oh for a
horse, a stock-whip, a lasso, or even a good collie! Wild West capers
are all very well, but the human actor must have the appropriate
outfit, or the whole thing may be a fiasco. And a fiasco it was. The
spirit went out of me, and I plodded wearily after them, until it
pleased them to check their mad career and turn back quietly to
the ford.

Here fresh trouble awaited me. The gate on the farther side
was shut. While I waded across to open it the cows, checked in
their advance, started a free fight in the shallow water, during
which the more timid dispersed. Not merely were these Guernseys
individualistic, and entirely without corporate spirit, but they
were deeply suspicious of each other, would not walk in a bunch
for fear of one another's horns, and refused to move freely through
water or any narrow place. In fact, they behaved exactly like the
armed nations of Europe, and the patient cowherd, as he watched
and controlled their silly capers, felt not only cross but sad, like a
disillusioned official of the League of Nations. The ford in any
case was a place of cursing and exasperation. Here cattle not only
fought, but instead of crossing, stood dreaming on the gravel-
bank in the middle of the stream. The bull stopped to rub his
neck on a fallen tree instead of passing through the gate; Jennifer
held up the whole procession to nibble at some bramble leaves
overhanging it. No sooner had some finished drinking than others
began. At last they were marshalled, and slowly we dawdled up
the road to the steading, shaded by Scots firs whose branches were
still festooned with trails of hay and corn brushed from the tops of
incoming harvest loads. Here the cows, unless blocked by some
visitor's stationary car, or deflected by an attack from the gander,
would go quietly into the byre, though no doubt into the wrong
stalls, so great is the power of perversity. In negotiating the ford,
the chief difficulty was the varying depth of the water. In a dry
spell, the river was not more than a few inches deep, and left in
the middle a long bank of gravel exposed, while after heavy rain

it would reach half way up a cow's flanks, and the driver be forced to cross by the footbridge lower down. When the stream was fairly full, the cowherd must wade at a snail's pace, and well back from the cows, or the water would slop over the tops of his gumboots; and to move in gumboots filled to the brim is a misery. The rise and fall was very rapid, making the depth at any given moment very hard to judge. Also, when putting a long string of cattle across, the first beast would be well into the steading when the driver was only just entering the water and quite unable to control anything on the farther shore. So it once happened that the bull, who had crossed at the head of the procession, caught sight of the forest officer ambling up the road on a cycle. He hated cycles, and with one frightful bellow started in pursuit. Leaving the cows to take care of themselves, I dashed after him, and was just in time to see the cyclist, bent low over the handlebars, and pedalling like a madman, fly through the boundary gate and slam it in the bull's face.

The river at Strathascaig drained a large area of low-lying flats, which in some cases were below the level of the stream and protected by banks. The land about its upper reaches, which belonged to the Forestry Commission, had been systematically drained, so that at times of flood an extra volume of water was discharged into a channel already too full to receive it. When a heavy downpour of rain coincided with a high spring tide, which for an hour or two dammed back the fresh water as it sought to escape to the sea, much of the Laird's land was flooded. Should there be any danger of this the cows, especially if they had calves with them, were kept on the near side of the river, for if the ford were impassable the only way home was a long round by the road. This circuit not only wasted time but spoilt tempers, for cattle homeward-bound detest having their heads turned away from the homestead, and are most unwilling to travel in a direction that in all appearance does not lead to fodder and stall.

On most farms those who are responsible for planning the position of field gates can never themselves have driven cattle or studied their natures. On winter afternoons, shortly before the customary time of coming in, cows always gather at the point nearest to the homestead, waiting to be let out of the field. The gate through which they have been put in and out a hundred times is somewhere else, probably at the opposite corner, but that makes no difference: they will always assemble at the point nearest as the crow flies to their byre, and can only be driven away from it by force. At Strathascaig, one of the fields most often used for the milk cows had its gate at the corner farthest from home, so that you must first drive them with blows and shouts up to the gate, keeping close beside the fence to prevent them from breaking back, then, after passing through, take a sharp right-about turn, and follow the same fence back again to the ford. Unless the cows were hustled through the gate in a bunch, the hindmost ones, seeing the leaders turn and walk down the outside of the fence, would turn themselves *before* passing through the gate, and follow the fence down on the *inside*. Confusion, fury, oaths and broken sticks were the sequel; and on one of these occasions I stumbled into an evil-smelling bog and sank nearly hip-deep. Though it was hardly a hundred yards from the farm, the Laird declared that he did not know of its existence. No: the gates of pasture fields should always be placed at the corner nearest home. But they never will be until every farmer is his own drover.

When Colin, the Laird's nephew, was a little boy, and being sent to fetch in the milk cows was rather a long time on the way, he would say by way of apology, 'But, Uncle, more than one cow is very hard to steer.' I agree; and more than one calf even harder. Calves have no sense of direction, and there is a certain type of stocky black or blue-grey beast that seems to be almost half-witted. They blunder to and fro, turn in at the wrong opening, start in terror at the most unforeseen objects, stop dead and refuse to budge an inch, and from the drover's point of view are the most exasperating creatures under the sun, except when shepherded by their dams. And even then they have an uncanny knack of getting separated from them. A calf can easily pass through a gap in a fence where its mother is unable to follow. And it is fairly safe to assume that if a calf, lamb, or chicken gets out through any unofficial opening it will not know how to get back by the same way.

At Strathascaig the Galloways and their calves were sometimes in the field next to our house. One night in October, just before the calves were to be weaned and sent away, I was roused by the bellowing of a cow, answered by the bawl of a calf, which from the direction of the sound I judged to be somewhere in our garden. I pulled on a few clothes and went out. It was bright moonlight, and I saw that my theory was correct. Somehow or other one of the Galloway calves had got over or through the stone dyke, and did not know how to get back. Nor could its mother reach it – hence the uproar. To and fro I chased the obstinate, bawling wretch, until more by luck than design it blundered through a gap at the boggy bottom of the brae and rejoined its distracted parent. Returning to the house, I saw the white bull standing at the gate. It was not the hour for an exhibition of bull-chasing, nor was I suitably dressed, so I left him where he was and went to bed. Next day I heard that he had escaped during the night, wandered up the road, snuffed at a hiker's tent, and meditated tossing it; but thinking better of this, had roamed off to a neighbouring farm, leaving the occupants of the tent half dead with terror. At the farm he was caught and brought back to his pasture. At what point in his wanderings he had arrived at my gate I neither know nor care. Later in the season he was to break in and uproot a rhododendron bush; but at that time he had never set foot within the sacred precincts of my mother's flower garden.

14

August: Making Hay While the Sun Does Not Shine

JULY was nearing its close, and the haymaking in full swing. We should now need an extra helper, and the choice lay between employing a local man daily, who though experienced would probably be lazy and certainly expensive, and a boy or girl of our own class to live in the house and work for keep. Good local labour was hard to find at this time of year, when all the crofters were busy with their own work. In 1933, when I had a girl friend as permanent assistant, I took a succession of Oxford women students to help with the haymaking and harvest. They were a merry party, working with great zeal, though of course without experience or very much quickness or muscular strength. This year, as Peter was already in the house, and also because I myself felt disinclined to do as much heavy forking and loading as I had done in the previous season, I advertised in *The Times* for an educated boy to work on a Highland farm in return for board and lodging. Among the replies was one from a minister's son, aged thirty-three, whose mother and sisters were willing to pay for his keep on a farm. Thinking that when a man of mature age is boarded out by his mother and sisters he must be either mentally deficient, or addicted to drink and drugs, or merely harmless to the verge of idiocy, I rejected his application and took Vivian, a public school boy of twenty-two, who had given up an engineering apprenticeship, and wanted some temporary work to fill in time till he could try his fortunes abroad. He had never worked on a farm before, and was unlucky enough to begin in a season which would have daunted the most resolute; but I do not think that he altogether regretted this experience of a Highland summer.

Many people refuse to employ voluntary labour. It is true that there is little to hold an unpaid worker, who may think that the absence of a money wage is a good excuse for slackness and irregular hours. But if one can secure the right people and maintain the right spirit, the voluntary system may work very well; and in my own case I have found little cause for grumbling beyond a certain amount of slowness, and in the case of the girls, a good deal too much chatter! Some people enjoy a spell of farm work in the summer, and I had learnt from the Laird the art of haymaking with amateur labour. He of course had more to offer in the way of amusements and good food; but the farm life itself, provided it does not last long enough to become monotonous, provides for many country-lovers a sufficiently entertaining holiday. But the farmer who decides to take voluntary helpers must be careful not to get himself landed with cranks, imbeciles, difficult young people, neurotics, or sheer abysmal fools. The country is certainly a place of healing and refreshment, free from most of the temptations and maladies of life in cities, but it is not a convenient dumping ground for the failures and misfits, a place where nervous invalids may discuss their complexes and super-annuated public school boys wonder how long it will be till dinner-time. We imagine that with the spread of a wider knowledge and a deeper understanding of other people's lives our old notion that the agricultural worker is naturally stupid, and his business therefore so elementary that anyone can do it, would have gone by the board, and we should recognise him as a specialist whose science takes a lifetime to acquire. Not a bit of it; parents who despair of finding a career for a son too sluggish or too feeble-minded to make one of his own, push or persuade him into agriculture, because in these days, when the Church is no longer a lucrative refuge for stupidity, they believe that a farmer's life makes no demands on intelligence. Nothing could be further from the truth. The number of things that a good farmer ought to know is so great, his qualifications so various, that the ordinary single-track mind becomes dazed at the mere thought of them. There is the old traditional knowledge – weather lore and land sense and that profound understanding of the habits of plants and animals that only comes with a lifetime of sympathetic observation. Then in these days, when agriculture has become so closely interwoven with other aspects of national life that it cannot be separated, the farmer must know something of finance, economics, and political thought, that he may understand the forces that govern prices and markets. Besides this, he should have some acquaintance with chemistry and biology in their practical application. He must also be able to handle his men and do their work as well or better than they do it themselves. Of course it is possible to farm, and to farm successfully, without all these qualifications; but the first-class farmer must possess a good many of them, enough at any rate to dispose of the notion that agriculture is only fit for fools.

On 7 August, after a wild exchange of crossed telegrams, Vivian arrived. Hearing that it rained often and heavily, he had brought a complete panoply of wet-weather clothes, and sorely he needed them. The day of his coming was fine, but as the months of August and September yielded an average of one dry day a week, it will be agreed that he was lucky. I much regret now that we kept no systematic weather log, for it would have proved beyond any possibility of doubt that from the middle of July till the end of October, we did not have twenty-four consecutive hours without rain, with the exception of one brief spell of two days in September, and even then our sense of security was broken by the sound of distant thunder! Nor except for a few hours now and then did the wind vary its monotonous round of SE–S–SW–W, and then back to SE with deluge renewed. We knew this sequence by heart. First came a pouring wet day with a strong southerly or south-easterly wind; or if the depression were slow-moving or stationary, with its centre somewhere near us, there would be no wind at all, but long relentless shafts of rain, which would soon penetrate the stoutest waterproof. Next a day of torrential showers with brief intervals and a wind still strong but veering a point or two to the west of south; and last, a succession of heavy rain or hail squalls, with longer and perhaps sunnier intervals than on the previous day and wind from a more westerly point. In a normal season this sequence would be followed by a longer or shorter period of dry north-westerly weather with a good breeze and varying amount of sunshine. But now these dry spells were omitted, and after the second showery day the barometer would fall, the wind back south or south-east, and all that weary round begin again. Every morning I would study the weather from my bedroom window, and could count on the fingers of one hand the days on which the hills of Skye were clear of rain. Even fewer were the days when the air was dry enough to evaporate the cluster of raindrops that used to decorate a tuft of rushes beneath the window. At that time Peter and I were milking together, and it was a matter for comment if we could not hear the pattering or drumming or thundering of rain upon the iron roof of the byre, or if we could feed the calves at the barrier without being drenched by far-blown spurts of water from a broken gutter-pipe.

All this would have been tolerable in winter, but when you must somehow get enough hay and corn secured to keep thirty-odd beasts in good condition from November to May, and play at summer in waterproof and gumboots, there may be just reason for complaint, especially when those rare fine days so often fall upon a Sunday.

In the remoter Highland districts, a Christian man's whole duty appears to be the observance of the pre-Christian Jewish Sabbath. Thus to avoid causing scandal one can only gnash one's teeth and watch the rare sunshine streaming upon empty fields and idle workers. If the hay on the fence at Strathascaig were dry on a Sunday, the Laird would wait till nightfall and clear it by stealth, a piece of hypocrisy that can have deceived no one, since fences observed to be full on Saturday night were visibly bare on Monday morning! On such occasions Colin and I would secretly fetch in the horse from the field, harness him with the stable doors closed, and steal into the open, covering the betraying whiteness of our shirts with dun waterproofs. The sky was overcast, and a puffy little wind, heralding rain, came out of the south, making the topmost layer of grass turn back upon the rope, like a girl's hair across her ribbon. In the dim light we could just see the loom of the loaded fence, but all details were lost: we worked by touch alone and at high speed. As we piled great armfuls of half-dried hay upon the sledge, we laughed discreetly and talked in conspiratorial undertones. The slats in the barn were closed, but through the chinks pale shafts of light streamed forth, revealing the presence of the Laird who, with a lamp propped on a barrel, was forking his unhallowed harvest. There were holy people walking on the road, who might see the light and hear the creak and jingle of the sledge. But it did not really matter, for we had not flaunted our ungodliness by daylight.

This year at Achnabo we did not get much chance of Sabbath-breaking, for on fine Sundays the beauty of the day was soon over. By four o'clock a fan-shaped veil of cirrus would be seen rising quickly from the southern horizon, spreading and thickening till it threatened the whole sky. As the diminishing band of blue retreated northwards, all brightness ebbed from the world; the sun's disc became wan and uncertain, then blurred by vapour to a luminous patch, and at last was no more than the unseen source of faint and rayless light. The breeze, which all day had blown freshly from the west, fell to an uneasy calm. The naked hills of Skye were dark against the whitish sky; above their peaks a few detached woolly clouds were floating like derelict ships upon a pallid sea; and as the veil behind dropped lower and denser, they lost all individual form and colour, and were merged in that invading pall. For a few minutes the jagged ridges of the Cuillins rose clear; then there was a distillation on their summits as of mist dissolving into water, and the sharp edge wavered, faded, and was gone. The nearer Red Hills were enveloped; they too vanished, and in front of the advancing rain a fitful wind arose, stirring the pines to wintry moaning. The horizons closed in, the sky above was lowered to meet them, and from this formless dome of vapour a few pattering drops began to fall. Half an hour later the pattering had become a deluge, driven in slanting pitiless shafts before a rising wind.

This pageant of approaching bad weather was to become painfully familiar to us, as also the infinitely various forms that rain itself can assume. In a town rain is just rain, varying only in duration or intensity. But the countryman, to whom it is one of the chief forces that moulds man's life, must look at it with more observant eyes. Rain, you will say, will have at least one constant quality — it must always be water falling from a cloud. Not it would seem quite always, for I have seen a heavy shower descend through a transparency of sunshine, or glimmer of starlight. Long ago at Strathascaig, a day of wind and rain ended in a calm night with a clear sky full of stars. But the stars had an unnatural look, being enlarged and misty at the edge; for all the time, though from no visible cloud, torrents of rain were falling. There is something disquieting about rain from an apparently clear sky; it is seen only in very dirty weather, as if some approaching horror were reversing the laws of Nature. We were not surprised to find that the barometer had dropped to 28°, and was still falling. We waited to see what would happen. It continued to fall until the needle touched 27·7°, the lowest reading that could be registered with this particular instrument. How much lower it may have fallen we shall never know. Nothing spectacular happened; but the memory of that strange and starlit downpour will always remain with me.

Of this and every other type of rain we had experience in our haymaking. There were misty showers that starred the feathery heads of meadow grasses with countless sparkling drops, showers that trailed soft nebulous veils across the distant hills, pierced with quivering gleams of sunshine, transient as the rainbows that often spanned them — a delight to the artist, but to the farmer at his work an unmitigated nuisance; for an apparently firm and anticyclonic cloud might without warning dissolve into a fine spray of rain, which bedewed the spread hay and made it impossible to carry. There was unbroken windless drizzle, with low banks of vapour drifting aimlessly among the hills, and if the rain should ease a little, down came the midges with vindictive fury. There were scattered spatterings from high clouds, and large detached plops in thunder plumps, and vicious lashings in sudden squalls, and long straight rods from racing scud, and horizontal shafts borne upon loud and piercing gales, with blasts that drove you quickly into shelter. Rain of all kinds and almost every day.

The crop was very heavy, green, and succulent, and the pressure of that eternal rain-laden south wind, with the sheer weight of the drenched heads themselves, made the grass lie over badly; it became increasingly difficult to cut, especially on the southward-facing slopes, where it was necessary to work the scythe uphill. Wet grass is easier to mow than dry, but one may readily have too much of a good thing, when continuous downpours flatten the crop to a mouldy carpet, and the stalks rot at the bottom for want of air and drainage. If Murdo and Peter had not been excellent scythemen, they would have made a sorry mess of the lower part of the field, which before cutting had been trampled by marauding sheep. But nothing daunted Murdo, who combined physical strength with the philosophic spirit that takes things as they come: if it rained moderately, he scythed in his shirt-sleeves, if heavily, he put on a coat with more holes in it than cloth; but if it poured beyond all measure, he would hang up his scythe in the cart-shed and saunter home, remarking that Achnabo was taking in hay while the experts were waiting for the fine weather. One trouble about this piecemeal and protracted haymaking was that if one waited for the grass to be ripe for mowing, a good deal of it would inevitably be past its best before it was cut. A friend from the south who inspected our hayfield in the middle of August remarked that most of it should have been cut long ago, and the hay would be of inferior quality. I showed him some of the new hay in the sawmill; he was struck by its greenness and fragrance, and admitted that it was worth £1 a ton more than the run of hay in the south, no doubt because the crop is dried by wind rather than by sun. It is true that early maturing grasses like cocksfoot and sweet vernal, in which our fields abound, have seeded and become rather tough by late August and September; but the prevailing moisture produces a thick undergrowth of succulent grass which gives the hay its excellent colour and perfume.

Our usual method was as follows. Peter and Murdo, sometimes Murdo alone, would cut as many swathes as could be conveniently dealt with in the next day or two. They would then fork the newly mown swathes into heaps, load it onto the sledge, and lay it along the wire fences ready for filling. Then some or all of us, as the occasion demanded, would fill the fences with grass. For the entertainment of large farmers in dry climates, I will describe this ridiculous but effective process in detail. You take up as large a handful of grass as you can hold, shake it, pull it out so that the stalks are lying roughly parallel, and lay it on the lowest wire, not too thickly, or there will be no through ventilation, and not too thinly, or else the grass, as it loses bulk in drying, will blow off. This is done, section by section, until all the wires are filled; the grass on the topmost wire (there are usually five or six) is tied down with lengths of rope fixed by a turn round each post and caught at intervals with ties of twisted stalks, and serves as a thatch to protect the rest from the wet. In good weather, that is to say, breezy but not necessarily rainless, a couple of days on the fence will dry the grass sufficiently to be finished on the rack in the barn. In dank or sodden calms, or if a gale has stripped the wires so that they have to be refilled, it may be six or seven days before the first drying process is complete, but meantime however fierce the deluge you may rest in peace. The crop, lifted from the ground and permeated by currents of fresh air, will not rot or spoil. The business of making hay on the fence is often criticised on the ground that it is slow and wasteful of labour. This depends altogether on the workers. With a little practice and judgement fence-filling can be done at high speed, and apart from refilling after a gale, the grass need be handled only once. It compares very favourably both in speed and efficiency with the constant spreading, turning, raking, and coiling required by the traditional method of drying on the ground. I have timed both processes on fine days, when the old method could be seen at its best, and the advantage was always with the fence. Once on the fence, the crop is safe at any stage, whatever the weather, whereas hay that is made on the ground is not secure till it is built into large coils, which can only be done in a dry spell. It is curious how few of the crofters care to avail themselves of this method, which will ensure good hay in the worst weather. It would pay them over and over again, even if they had to erect temporary fences for the purpose. At Strathascaig, fence-filling has given uniformly good results for over twenty years, even in the notoriously bad summers of 1923 and 1924. And in 1934, the worst haymaking season within living memory, it has been our only salvation. No doubt the idea originated in Scandinavia, where various crop-drying devices are in use, but it has certainly proved its value in the Highlands, and might well be tried in other districts where farms are small and rainfall heavy.

In the last eight years I must have filled many miles of fence, and should by this time know all there is to know about it. When the weather is dry and breezy enough to banish midges, and you have pleasant company or agreeable thoughts, fence-filling is by no means a bad job. But when you are alone and full of melancholy, when the rain soaks through the shoulders of your waterproof and drips from your streaming hair into the hollow of your collar, when the sodden grass is as heavy as lead, and clouds of persistent midges raise large white lumps on every exposed inch of skin, then the fencer's lot is indeed a hard one. The heaps of grass waiting to be dealt with are of Himalayan stature, and the wires yet to be filled, like all other parallel lines, could be produced to infinity. Confronted with this task, which must at first, if not always, have seemed useless and tedious, our voluntary workers did very well. They were slow, they may have been tired or bored, but they never grumbled and somehow the work got done. I have noticed with amusement that when new at this job a boy will nearly always put on the grass too thick, and a girl too thin. The correct thickness is a matter of judgement, depending largely on the strength of the wind and quality of the grass; but if the bundles are put on with stalks more or less lying the same way, as in a sheaf of corn, it is surprising how much a single wire will carry and yet admit enough air for rapid drying. The one fatal thing is to cram on the grass anyhow; more haste less speed. It will never dry until it is all pulled off the fence, shaken out, and replaced.

Fence-filling is best done in pairs, each one taking his own section (i.e. the four- or five-foot stretch of wire between two posts) and roping as he goes. There is rivalry in completing the sections, and the practised fencer has plenty of spare energy for conversation without any loss of speed. When Colin and I filled fences at Strathascaig, we discussed many subjects – banking, money, divorce, and what is meant by life. I can see him now, a well-grown youngster of fourteen with a mop of dark hair, grey eyes, and a snub nose sprinkled with big freckles; his ancient grey knickers, which were getting too short for him, revealed a great expanse of scratched and sunburnt leg. Pausing for a moment in his work, he opened a new topic, to which, being a Scot, he had given a shrewdly practical turn. When he grew up, would it be better and cheaper to keep a housekeeper or a wife? The housekeeper must have wages and would probably waste the stores, while the wife might have children and cause other unforeseen expenses, and . . . At this point the Laird was seen striding across the field, his kilt swinging, his pockets bulging with oddments of possible use, his tacketed soles turned up at ever step. Colin began to work furiously with his eye on the approaching figure. Is Uncle in good tune to-day, or is he in a rage? Impossible to tell, for his face was as expressionless as that of a crack bridge-player. 'Work,

work, work, for Uncle's coming!' chanted Colin in the raucous tones of one who hopes to hasten the breaking of his voice.

The girl haymakers at Achnabo had Colin's love of abstract discussions, but they argued at greater length and had not yet learnt the art of talking without stopping work. They twittered away like birds, three or four at one section, discussing dramatic values or the theory of beauty, with little bundles of hay in their hands, until a shower or the coming of the lady of the harvest reminded them that they were engaged in agriculture.

For carting the grass to the fence when green, and from the fence to the barn when semi-dry, we used a wooden sledge, which was easier and quicker to load than a cart and (as we had no cart-frame) would carry nearly as much. The sledge consisted of two strong runners about nine feet long and three feet apart, with a strong cross-bar in front, and six or seven lighter boards behind, projecting about six inches beyond the runners on each side. The swingle-tree was fixed to a loop of fencing wire which passed through holes bored in the front of each runner. This sledge was simple to make, and luckily for us, equally simple to repair, for at first it suffered almost daily damage. Dick disliked sledging, and whenever possible would smash the boards by stepping back upon them, or cut a corner fine and run the sledge against a post, upsetting the load and breaking a board or two in the process. With exasperating intelligence he would always choose to do this when a load of dry hay was being rushed in before an advancing shower, and as most of the field gates were set at awkward angles, he had no lack of opportunity. Whenever possible he would get his feet entangled in the traces or reins, and when finally extricated with much swearing and loss of time, would either refuse to pull the load at all, or start with a jerk that would break the wire loop. More than once he bolted with an empty or partly loaded sledge, and in the pursuit would fall foul of a gatepost and smash a few boards. On one occasion one of the runners was broken, but we repaired it by stapling on a piece of strong wire, and for many days the sledge held together, but it had become curiously supple and followed the sinuosities of the ground like a slowly moving snake.

Loading the sledge with succulent rain-drenched grass, and goading an unwilling and bad-tempered horse was heavy work, mostly performed by Murdo and Peter, while Vivian, Mr Gordon, and I filled the 400-odd yards of five-wire fences, which when completely full, as we tried to keep them, held an unexpectedly large amount of hay. Stripping the fence was a quicker and pleasanter job, but the grass, which by this time had lost two-thirds or even three-quarters of its original bulk, had to be packed carefully on a broad base with someone stamping the load, and then lying on top to steady it on the journey home. Alec, who fancied himself with the ladies, would always induce some fair girl haymaker to stamp his load; but in 1934, when I was the only woman in the field, the work was organised on less romantic lines, and was, I fancy, rather more quickly done.

It needed some judgement to decide when to strip the half-dry grass from the fence and cart it to the racks indoors. If it came in too soon, the barn and sawmill became choked with too much damp material, which neither dried itself nor allowed anything else to dry. If it were left too long, there was much risk of getting it blown off by a gale or re-wetted by a fresh downpour, and until some sections of the fence were cleared there was little use in cutting more. As the season advanced and the grass in itself became weathered and therefore drier, the time required on the fence grew less, especially as there was nearly always a strong wind. Indeed the continual wind alone saved the crop, for had we suffered from similar deluges in calm weather, not a hundred-weight of hay could have been secured.

At times, however, we had too much wind. During the haymaking season there were four whole gales – on 21 August, 3 September, 8 September, and 22 October – besides many minor blows. On these four days the wind rose to hurricane force, and as luck would have it the grass on the fences had been there for a day or two and was dry enough to carry, had it been possible to handle it. But we could only stand idle and watch the wind strip every blade from the wires and whirl it far to leeward. Much of it rose high in the air and sailed away into the oak woods, some vanished completely, and a little was left on the ground beside the fence. This, when soaked by the following deluge, had to be raked up and put on afresh. We had hardly finished repairing the damage done by the gale of 3 September when the storm that came five days later involved us in further destruction. The hurricane of 22 October, which was accompanied by torrential rain, wrought unprecedented ruin. Every fence was stripped, large coils, including one that was almost as big as a stack, were overturned, and reduced to sodden and shapeless lumps. With this havoc Peter had to cope almost unaided, for Murdo had been paid off at the end of September, Vivian's time was up, and I, believing that by the middle of October the hay would all be finished, had gone for my usual after-harvest holiday. Rattray's shepherd, who was without a barn, and could never get a day dry enough to build a permanent stack, had all his coils blown over and half the crop ruined. This storm was followed by a period of torrential rain and showers of hail, which on the 31st culminated in a heavy fall of snow. Thus ended the last month of the hay, which yielded a total rainfall of over twelve inches.

We had, as I have said, about 400 yards of drying fences, and with the eternal wind we could count on a fairly steady flow of half-dry grass to the racks indoors. One week, to the amazement of our neighbours, we took in a sledge-load of good stuff every day, often snatched between heavy downpours, for if there is a strong breeze the grass, with the exception perhaps of the top layer, will recover from the worst wetting in a couple of hours. Our chief difficulty was the lack of drying space indoors. We could have done with racks twice the size, and to relieve the congestion in the barn and sawmill, we built some of the grass stripped from the fences into coils, the smaller ones solid and the larger ones centred round tripods, with an air space in the middle. A coil is, of course, a stack in miniature; as a bun to a cake, so is a coil to a stack. It is small, temporary, and immensely useful; and if well built, will stand without much damage for three or four weeks. The making of a good coil is a work of art. The best ones are thimble-shaped, with a base broad enough to weather a gale, and sides steep enough to throw off the rain. It takes much experience to estimate the size of base required for a given quantity of hay, and the whole structure must reach a height sufficient to allow for settling. The top, which should be well pointed, is secured by a rope lying in the direction of any expected wind, and the sides combed down with a rake to resemble a thatch. A perfect coil can be made alone only in comparatively calm weather; if there is much wind, an assistant is needed, and even so the hay can only be scrambled together anyhow on the chance of being able to carry it next day or rebuild. If the grass is on the green side, a dozen or so of small coils may be built; next day they are if possible spread in the sun and wind and remade into half the number; and later on these six can be combined into two or even one. We found that by using the tripods and corrugated iron ventilating tunnels we could build a large coil of half-dry grass which would neither heat nor rot; but even these hollow coils could not be taken in without a preliminary spreading on the ground. And so brief and uncertain were the dry intervals, so wringing wet the ground, that we hardly ever ventured to spread more than a few forkfuls at a time. Luckily for us, the horizons were wide and open, and we got early warning of approaching showers; but even so, there was need for great alertness, as the most unsuspicious-looking cloud would let fall a deluge, while a black and threatening sky might produce nothing but damns and confusion. One reasonably fine day, towards noon, we had taken to pieces a large coil and spread it thinly on the ground, so that it covered a large area. No rain cloud threatened our peace, and I had gone in to hurry on Flora with the dinner, leaving Peter and Murdo to turn the hay and rake it into windrows. Glancing casually out of the scullery window I saw that the hills of Skye had vanished. A heavy shower in the offing, and it would take half an hour or more to rake that hay together and build it into secure coils. I supposed that the boys had seen the shower and taken appropriate action, but ran out to make sure; and there in the middle of a sea of rippling windrows were the two beauties with their rakes, gazing at the sky and discussing whether it was going to rain or not, as if there were any answer to that question but one in this summer of 1934! I yelled to them and seized a fork myself, and we coiled everything in record time. Peter kept on arguing that the shower would go elsewhere. Actually it came our way, but was not very heavy. On most occasions we did not dare to spread the coils at all, with the result that they were rather damp when they came in and had to be dried on the racks, whereas spread coils could usually be forked straight back to the permanent pack at the end of the barn.

Besides the work in the field there was much forking to be done in the barn and sawmill. When a load came in from the fence it was shaken out and spread on the rack, thickly or thinly according to its own condition and the strength and drying qualities of the wind. Here it would remain until thoroughly dry, which might be two or three hours, a day, two days, or even up to four days in damp and windless weather. It was then forked on to a pile known as 'semi-finals', where it was allowed to mature a few days before being packed back for good. Its place was taken by a fresh load from the field, and the process begun anew. It will be seen that all this detailed manipulation of hay takes time, care, and judgement, a perpetual testing, grading, shifting, and handling which alone ensures success in such a climate. The stage at which judgement is most difficult is the final packing. In spite of all precautions the large pack in the sawmill was discovered to be slightly heating, and had to be opened and thrown loosely on the floor, where it was left to ventilate for some days before repacking. During this time no fresh loads could be taken to the sawmill, and the grass from the fences had to be made into coils.

The critic will wonder if the hay we get is worth so much

trouble, and whether it would not be simpler and cheaper to buy in hay from the east coast, or use some other feeding. I think on the whole it is. There is a glut of grazing in the summer, and the ten or twelve acres we shut up for mowing are more useful as a source of winter food than as extra pasture when grass is plentiful. I reckoned that our hay, counting Murdo's wages and the keep of our voluntary workers, cost about £1 per head of cattle for the season, that is, the price of 2 cwts of linseed cake or of 3 cwts of bruised oats. Bought-in hay is dear and of inferior quality, and must be supplemented with other feeding, whereas it is agreed that good meadow hay is the most nutritious and palatable food for all classes of cattle. We have to do something in the summer, and it may as well be haymaking as anything else. The slowness of the process is its real disadvantage, and if it were possible to devise an artificial drying plant on a smaller scale than those at present on the market, a larger area could be cut at one time, either with a mower or with several scythemen, and the hay be secured in a third of the time. But with the West Highland weather as it is, I challenge anyone to make more hay, or to make it quicker or better by any other method. We began cutting on 12 July, and took in the last load on 9 November. There were, as I have said, twelve inches of rain in October, while September, August, and the latter half of July were very nearly if not equally wet, so that the whole period must have yielded between thirty-five and forty inches, or the total annual rainfall of one of the English south-western counties! Of our neighbours, only those who started early and worked hard secured a proper crop. In many parts of Skye the hay was never cut at all, or spoilt in the field, and large quantities have been imported by rail and steamer. There is only one way to security: begin cutting as early as possible, and when once started, carry on regardless of weather. It is useless to wait for a fine spell that may never come, and find in the end that your grass is too rotten and badly laid to be worth cutting at all. Carry on, and *put everything on the fence*; grass can be scythed, carted, and fenced in the worst deluge. Out of all the bad days we had, there were no more than four or five (Sundays excluded) when we did not work at the hay, and these were days of excessive wind, when fencing was physically impossible. Thank God it is all finished now, and if we are lazy in winter, I think we deserve it!

15

Interlude: In Quest of Food

IT was during the haymaking season that I was most acutely conscious of the difficulty of being both farmer and farmer's wife. The Laird, I reflected, can work cheerfully at his hay in the certain knowledge that good meals and comfort in the house will be provided, or at any rate superintended, by Herself; while I must ever be keeping one eye on the weather and the other on some boiling pot. The first summer I had no servant, and did all the cooking, washing, and baking for five people, as well as most of the housework, all catering and supervision of stores, and as much haymaking as I could fit into a sixteen-hour day. This was rather too much of a good thing, and the second summer I employed Flora, who was engaged to clean, cook, and wash for the household, while I carried on with my work as a farmer. I had better confess at once that though not quite as helpless as many women of my undomestic antecedents, I am but a poor housewife. I can cook moderately, and bake a good loaf of bread, but house-cleaning of any description is abominable to me. I delude myself with the idea that a house in the country does not get really dirty, and that a lick and a promise is as good as a spring-clean, except in the kitchen, where I have rather a weakness for well-swept floors and spotless sinks. In the West, time and order do not seem to matter very much, unless we want to catch a train or find something. The alarm clock on the kitchen mantelpiece, if we do not forget to wind it and no one jars the regulator, keeps something which, if not exactly GMT, serves to get up and milk the cows by. If anyone wishes to catch a train we can ask the correct time from a neighbour with the wireless. Or we can consult Mr Gordon, who though meticulously careful, has been known to let his watch get twenty minutes slow. In January, when it is pitch-dark at seven, I keep our clock fast on purpose to help us to get up. As for order, my own theory is elimination. Have very little in the house, so that it cannot get beyond a certain limit of untidiness; thus it is possible to hunt through everything in an hour. Keep also some boxroom or capacious cupboard into which the more respectable rooms can be tidied. Thus like the Roman conquerors of Britain we make a solitude and call it peace. In any case why bother? If, as scientists keep on telling us, the random element in the universe is continually increasing, tidying up is a hopeless as well as an unscientific business. So that neither Flora nor I should be blamed for instinctive obedience to a universal law.

The great domestic event was the cleaning of the flues, which took place every Saturday. In building a modern range into an old-fashioned chimney, the plumbers had somehow left too small a space at the back, so that the smoke got away too slowly and left a large amount of soot behind. The chimney was periodically cleaned by Murdo, who climbed onto the roof and let down on a rope a heavy stone, to which a thick bunch of heather was attached. This primitive sweep's brush was quite as efficient as the modern variety and much quicker and cheaper. When Murdo had finished, his appearance was strictly professional, and he once mortally offended Alec by going into his cottage, using his sink to wash at, and drying himself on his towel. The real grievance was that he had gone home without cleaning the sink or rinsing the towel, a complaint which somehow did not ring true, as both sink and towel, though not actually inky black, were very far from clean when Murdo found them. High words ensued, and the couple were not on speaking terms for some days.

Sweeping the chimney was, however, a much less messy business than cleaning the flues. On Saturday morning Flora would set the alarm for half an hour before the usual time, and descend to the kitchen with a grimly resolute scowl. Here she would bang

and rattle with rings, bars, and fire irons, whirl her flexible wire
brush, rake, shovel, push, and pull, all to the accompaniment of
loud crashing noises and blinding clouds of soot and ashes. Half a
pailful of soot would be carried outside, the rest would rise in the
air and then slowly settle upon furniture, mantelpiece, and floor,
until the whole place looked as if a volcano had recently erupted
in it. Soot lay powdered on the top of the clock, on the glasses of
pictures, on the surface of Bibby's calendar; it outlined the shapes
of cobweb festoons, it hid the pattern of the linoleum. Donas,
who used to lie on the hearth, had the white parts of his body
turned to grey. In the midst of this desolation stood Flora, with a
stove-brush and tin of Zebo in her hand, and on her face the
marks of recent toil. Her eyes were straying towards a sheet of
The Times spread on the table to protect it; when I had gone out
to milk she would settle down to read while waiting for the
porridge water to boil. She was a great scholar, and the intervals
of cooking were always filled by miscellaneous reading on the
kitchen table. When Flora left us, Peter took over the flues, for
which service, cleanly and quickly performed, I hope he will go
to heaven. In a recently published life of Wilson of the Antarctic,
I read with delight that after his marriage he always cleaned the
flues. That was indeed the ideal husband, and I as a mere spinster
am lucky to find anyone to do it for me.

Food was a more serious problem. There were five hungry
people to feed, not much money to spare, no shops at hand to
avert a sudden crisis. Bread and porridge were secure, for the
kitchen girnel held a boll of flour and another of oatmeal; early
potatoes could be dug from the field, and we had plenty of milk.
But eggs, butter, meat, and jam were all uncertain and worrying.
Our poultry stock was all young, and the pullets would not start
laying till the autumn. We had to ransack the village for eggs, or
drag basketfuls home from Strathascaig on Sundays; they were my
chief resource for tea, and without them I felt as helpless as a ship
without a rudder. Butter we no longer made at home, and though
the baker was supposed to bring the New Zealand variety every

week, he often forgot to do so or did not come at all, and some obliging person had to post off to a shop nearly three miles away to buy a fresh supply. Meat had to be ordered by post. The arrival of the butcher's parcel was a great event for the dogs, who gazed at the torn and bloodstained paper with rapt adoration. Unwilling vegetarians, they were, like the shepherd's collies, fed mostly on porridge, and rarely had meat unless they killed a rabbit or (heaven forbid!) discovered a dead sheep on the hill. Even bones did not often come their way, since it was more economical to buy cuts with little or no bone. For this reason I rarely bought the mutton of the local butchers, but sent to Inverness for beefsteak, silverside, or a whole tongue. In summer, the problem of how to get enough to last for several days without any of it going bad used to keep me awake at nights. I mostly solved it by having salt silverside one week and pickled tongue the next. Luckily we were all too hungry to kick at the monotony, which was only varied by a weekly cockerel of our own rearing. A few rabbits were shot, but no one cared for them, and they hardly seemed worth the trouble of skinning.

I must confess with shame that apart from a few pounds from the blackcurrants in the steading, no jam was made at Achnabo. We had no other home-grown fruit, and Flora never seemed to have any time for extra work – I certainly had none. I fully intended to make some bramble jam, and even bought half a hundredweight of sugar for the purpose; but owing to the incessant wet weather very few brambles ripened, and those that did were tasteless and mildewed. This was a great disappointment, for at Strathascaig we had always been able to gather large quantities of them, making use of the still weather so common in September, when nothing could be done to the hay till late afternoon.

I remember one grey, windless day in late September. A pall of ashen cloud, softer than sleep, covered the sky and all the earth above 1,000 feet: its straight clear edge lay close against the hills. The loch gleamed pale and metallic; the slopes beneath the clouds were deepest purple, with here and there faint tawny blooms where shafts of light struck down into some corrie. The brilliant green of the aftermath, and splashes of red and gold among the fern and birches lent colour to a scene which would otherwise, especially in the distinctness of detail, have recalled the concentrated black and white of an etching. The day would be good – the straight cloud-edge and luminous shades beneath were signs of that. But the dew had been very heavy, and unless a breeze got up we could not handle the hay.

Good: for Colin and I had other fish to fry. It was Monday morning. No one was so wicked as to go brambling on Sunday, but Nature was doing her work as usual, so that the alluring berries would have had a quiet day in which to ripen undisturbed. It was early too, and we need not dread those armies of determined women with ancient waterproofs and pails, who marched for miles along the line to spoil our bramble thickets. Not really ours: I believe they were the property of the LMS Railway, and that all trespassers, we as well as they, were liable to penalties. But these meaningless regulations, like the notice-boards which once proclaimed them, had long since crumbled into dust. Clothed in ancient waterproofs and superannuated gumboots, with quart milk-pails suspended like sporrans from our waists, we made for the railway, each carrying a two-gallon pail into which the small tins could be emptied. We scanned the line, not indeed for trains, for who cared about anything of such passing interest, but for rival bramblers. The polished curve of the rails, the tarry sleepers just too far apart for comfort in walking, the trodden gravel at the edge, stretched from our feet to the shadows of the next cutting, empty of life, and in their stillness more suggestive of a well-preserved Roman aqueduct than of anything connected with modern transport. Good. We should have the field to ourselves.

In most tales of travel we read that at some point the hero was compelled to subsist upon berries, which doubtless saved his life. And yet the gathering of berries, however arduous and fraught with danger, is considered beneath the dignity of an adult male. A curious convention, like that which decrees that no man who is

217

not a sailor may sew on a button, so that the Laird's shepherd proclaimed his widowed state by lashing on his coat with rope. It is right for a man to sit all day on a camp stool beside a pond, or to blaze away at a dress parade of tame pheasants. That is sport. But to struggle in a thorny thicket for a dozen pounds of jam, full of vitamins and delight, is not sport. It is woman's sphere.

Now brambling, if regarded in the right light, might provide just that outlet for the combative instinct which in our sedentary and rather mawkish civilisation is very hard to find. Especially brambling on a railway embankment. Brambles are everywhere full of thorns and long entangling suckers; but nowhere are they as formidable as on the steep sides of the line, where loose stones and crumbling gravel give little foothold to the picker. The shifting soil is covered with a dense growth of plants and bushes, every one of them a painful and prickly impediment. Whins, wild raspberries, briars, thistles, nettles – all these abound, and a brambling expedition is one long falling among thorns. Full of the lust of battle, we climbed the railway fence, setting the wires jangling from one side of the glen to the other, and plunged into the thicket. Oh, joy! The stationmaster's fat and gallant wife had by sheer weight beaten a path before us like the track of a hippopotamus in the reeds. She must have been there on Saturday, and her methodical picking would have diminished our bag; but the Sunday truce had ripened a good many more. We worked furiously, fighting for a foothold on the shifting gravel, beating back the thorny embrace of barren suckers, reaching for the highest berries and often over-reaching, so that we were only saved from falling by an agonised clutch at the nearest whin-bush or trail of briar. Berries rattled into the tins; our mouths and hands were deeply stained with purple, and on the unprotected strip between gumboot and waterproof, blood trickled down from a thousand scratches. Beyond the cutting, a puff of smoke appeared, and in a moment the still air was full of noise. The train came grinding round the curve, laden with tourists returning from Skye and Lewis. Faces were pressed against the windows, and Colin, full of adolescent self-consciousness, was sure that they must all be gazing at his scratched and dirty knees. After this burst of publicity we resumed work, until after many emptyings of the small tins, the large pails were full.

This railway line is probably the most beautiful in Britain, shadowed by woods which extend for nearly three miles along the northern face of the Strathascaig crags. The wooded belt is very narrow, covering a steep slope from the level of the loch for about 300 feet, whence the rock rises in a sheer cliff to a height of 1,000 feet above the sea. Beside the loch is the line; parallel to it, and some distance above it is the road. Both road and railway are overhung with trees which, planted more than sixty years ago, are some of them of great size and of many varieties – fir, spruce, larch, ash, yew, laurel, rhododendron. For many years the wood has received no attention; in some places the younger trees are too dense, in others the ground is cumbered with fallen and rotting giants, whose interlocked branches and twisted roots have produced an inextricable tangle. Much timber there is going to waste, for it is worth no one's while to cut and remove it. The best view of the wood is from a boat on the loch, and the romantically minded person will be surprised to find that the railway only enhances the peculiar beauty of the place. Numerous cuttings form a series of picturesque little cliffs, fringed with trees and adorned with ferns and heather. In some places the embankment crosses a bay, enclosing a still, semicircular pool, into which the tide is poured through a narrow bridge until it brims to the edge of the wood, and then ebbs to leave it almost dry. At one time these pools were much frequented by cuddies, and a crofter had the bright notion of closing the exit with a net, so that at low water the muddy bottom was covered with stranded and dying fish. The best time to visit the pools is at evening, when a good breeze is blowing outside. No sooner has the boat, with hollow roll of oars, passed under the bridge, than the eager jobbling noise of waves is still; no sound is heard but the tinkle of a burn and the soughing of the wind in the pines. The dark semicircle of trees, crowding, secret, and already full of coming night, has every detail reflected in the placid water of the pool. We rest upon our oars; the boat, moving with the tide, steals on towards the shore. There, where a little burn flows out into the loch, we will find a landing-place. We might be entering the lagoon of some uncharted South Sea island. A hoarse whistle sounds; the evening train, a collection of miscellaneous trucks with a passenger coach attached as an afterthought, comes puffing round the bend. Beside the line, a black-coated old man with a long white beard stands watching. A newspaper is thrown to him, and he remains fixed in the middle of the track, absorbed in his reading. The train is out of sight, and he knows there will not be another till tomorrow. He is the missionary; his hens roost in the branches of that magic wood, and the homely flapping of his weekly wash drowns its mysterious murmurs.

In all this railway forest there is one tree that has a special fascination. It is a young sycamore, which grows on the edge of a cutting, closely pressed by tall pines. The near presence of other trees has made it long in the stem, with an airy tuft of leaves and branches at the top. In summer it is scarcely noticed, but in spring and autumn its intense young green or vivid orange stands out against the dark unchanging background of its neighbours. Stirred by the wind, it sways like a dancer. For a week or two in late April, and another week or so in October, it is the loveliest thing in the wood.

Apart from its beauty, the railway serves various local purposes which perhaps were not originally contemplated by the company. The gravel edge beside the rails, when not too recently repaired, makes an excellent level track for cyclists and walkers, and is often the shortest and easiest road between two villages. Trains are infrequent and their times well known, though one can some-times be caught unawares by specials or delayed goods trains, whose approach, if there is much wind in the trees or the loch is rough, cannot always be heard. To be caught in one of the numerous cuttings is to the novice an alarming experience; but the trains never go fast enough to cause much displacement of air, and there is plenty of room to stand back against the rock and watch the juggernaut roll by. In the early days of the railway it was possible to make a good living by driving sheep and cattle on to the line and claiming compensation. But this belongs to the golden past, when drivers would wait for twenty minutes at a wayside station to see if that distant speck upon the road would prove to be someone wanting the train. Now they run punctually enough to set a Highland clock by, and the local utility of the railway, besides serving as a substitute for a road, is confined to supplying us officially with old sleepers for sheds and foot-bridges, and unofficially with those iron 'chairs' which make such excellent anchors for boats and trammel nets.

The romantic reader will object to so much enthusiasm for the railway, which must, he thinks, defile the solemn beauty of its surroundings. As a matter of fact it does no such thing. I cannot tell you why: unless it is because the railway is not the creation of this age of noise and vulgarity, but dates back to the time when roads were alive with the glory of horses and the sea with the grace of sail. Or it may be because a train must run on a fixed track, and can easily be avoided. At any rate there are few people who would feel that a single-line railway with half a dozen trains in the day was any real menace to natural beauty.

16

September: Harvesting in the Deluge

WE had, as I have already said, nine and a half acres of corn. The first sheaf was cut on 27 August, and the last came into the sawmill on 13 October. That it should have taken seven weeks to secure so small a crop is a measure of the bad weather we had to contend with. There were six of us – three scythemen and three other workers and given a dry week we could have got both fields cut, bound, and stooked by the end of it, even without machinery. As it was, the long field took a fortnight (27 August–10 September), and the field below Rattray's five days (10 September–15 September). The following four and a half weeks (15 September–17 October) were spent in frantic attempts to get the sheaves dry enough for stacking, and our persistent efforts were rewarded by securing about 95 per cent of the crop in good condition. This was achieved only by working early and late, and handling every single sheaf of about 6,000 *individually*. In no other way, I believe, could so much corn have been saved, but I do not want to do it again. It will be remembered that we had the hay on our hands at the same time; but the corn had the priority, and we made hay only when it was too wet to work in the cornfields.

I had sown White Line oats, a variety specially bred to be stiffer in the straw and less liable to lodge in bad weather. It justified its reputation, for though both fields had been fallow the previous season and yielded a long-strawed and heavy crop, the corn stood up gallantly, even to the hurricane and deluge of 21 August, which caught it nearly ripe and on an exposed slope. As our oats are used for feeding in the sheaf and not for threshing, it is advisable to start cutting when the crop is on the green side, especially in such a climate, where long delayed reaping would allow the parts left last to get over-ripe. So, when the deep green field was slightly tinged with yellow, and a sunny morning promised a good day, we decided to make a beginning. At dawn the hay on all the fences was bone-dry, and we took in six large loads before leaving the hayfield and adjusting our scythes for the corn. We had barely worked for three hours when it began to rain; the sheaves were hastily stooked up and we retired to the barn to fork hay. Nothing further could be done till the 31st, and then work was once again stopped by rain in the afternoon, and haymaking resumed. The following day was fair in the morning, but nothing could be done after dinner, nor for the next two days at all. The 4th, 5th and 6th of September were fine, and we were able to work all day; by the last evening only a small patch was left, but owing to renewed downpours this could not be cut till the 10th. The short spell of good weather had helped us forward, but as luck would have it, Peter developed a bad cold on the 4th, and on this and the following day we were one scythe short.

In addition to Murdo, I had engaged the shepherd's nephew for a few days. Both these lads were experienced scythemen, and Peter by this time had become proficient. Vivian, Mr Gordon, and I went behind to bind the sheaves, and when enough had been cut the scythemen joined us in binding and stooking. In so treacherous a climate it is risky to leave corn on the ground, and I made a practice of having everything stooked up as we went along, so that if a shower came unexpectedly we had nothing to worry about.

After a good deal of practice at Strathascaig, I had become a fairly rapid binder, and my share of the harvest work at Achnabo was mainly the tying of innumerable sheaves. To the spectator at the edge of the field, this seems a simple, and, if he is romantically minded, a beautiful task: you just gather a bundle of corn, lift it, dandle it like a baby, pull out two or three straws to form a tie, bind it, throw it aside, and gather another one. But to the devil with romance! Sheaf-binding is about the worst and most back-breaking job on the farm, and, like so many other rural operations, far less simple than it looks, even when the swathes are lying evenly; if they are in disorder, as often happens when the corn is laid or unskilfully cut, it is almost impossible to make a success of it. If the sheaves are too big, they will not dry; if too small, they will not stand up. If the butts are uneven, the stooks will be crooked, knock-kneed, and damp at the base. If the band is too tight, the straw under it will not dry; if too loose, it will slip when the sheaf contracts in drying; if unskilfully made it will come untied when the stooks are forked into the cart. Even if the crop is clean, the harsh straw frays the skin, so that a day's work leaves the hands and arms covered with a network of minute scratches. And when hemp-nettle is present, which combines the prick of a thistle with the sting of a nettle, the binder's lot is indeed a hard one, especially as prolonged stooping gives him a frightful back-ache. At Strathascaig I used to think that hemp-nettles were the worst weeds in corn, but since coming to Achnabo I have changed my mind. The top of the long cornfield, in which no cleaning crop had been grown since I know not when, was thick with tall bristling thistles. The sheaves could hardly be handled without gloves, and the speed of binding was slowed down by half. The stooker has the best of it, but even his task is a thankless one if sheaves are as badly made as most of ours. He walks hither and thither, picking up sheaves and setting them up in fours if the weather is quiet and settled, with butts firmly straddled out and heads together, so that the air can circulate round the base. In a high wind, ten or twelves sheaves must go to the stook: they are rapidly hurled together, embraced, and securely tied.

The sun is low in the west; the last sheaf is bound, the last stook built. The field, resembling a vast encampment of tents, is left to providence. Looking back from the east, the tufted heads of the stooks are seen ringed with golden halos, like those that glorify sheep or any hairy beasts that stand against a level sun. So far good; but now comes the real struggle. The corn must be left eight or ten days in the stook to mature before it can be carried, and this interval is full of risk. Many a time the deep blue of the sea will change to angry white and steel grey; squalls sweep the field, and half the stooks are down. The rest, built of ragged amateurish sheaves, stagger like drunken men. The wind drops, and rain soaks down upon the fallen wreckage. The field has still its martial suggestion, but now it seems no longer full of tents, but of the wounded and dying survivors of a battle. We go out like Red Cross workers to raise the fallen. But a stook rebuilt never sits as well: the sheaves are flattened and drooping, the whole erection knock-kneed and ready to fall again.

The stooks in the long cornfield, which lay exposed to the full force of a south-westerly gale, had a severe battering. Nearly all of those made up to date were blown down by the storm of Monday the 3rd. They were set up again the same afternoon, and stood fairly well until the even worse hurricane of Saturday the 8th destroyed the whole lot, including all those that had been erected in the interval. This day was memorable. The whole field, with the exception of the small patch mentioned above, was in the stook. We had gone out after breakfast to reset a few that had collapsed in the night, when it began to blow. The wind rose rapidly; stooks fell faster than they could be set up, and in half an hour the whole field was flat. The sky was full of rain, which would most certainly fall as soon as the wind moderated. Much of the corn had from necessity been cut wet; it had been soaked in the stook by a whole day's downpour on the 7th, and would not benefit by lying drenched on the ground over the week-end. Realising that the stooks would never stand unsupported, we rushed out three or four of the tripods and a quantity of stack rope. In ordinary weather the sheaves would have been far too damp to build into hand-ricks, but we reckoned that so strong a wind would blow right through the interstices of the newly built and as yet unsettled mass, and dry the corn better than would be possible if it were lying exposed on the ground. In this we were right, but the difficulty was to keep the erection from blowing

225

away in its early stages. We spread the legs of the tripod as widely as possible, and then anchored it on the windward side by a rope stay passed from the top of the tripod to a peg driven into the ground. Mr Gordon and Vivian gathered, while Peter built and I held the sheaves in position. When the hand-rick reached waist height the stay was no longer required, but we had to build and rope the top with the help of a step ladder. The wind roared across the field and groaned in the pinewood behind it; tall trees lashed to and fro like saplings, while far below the hay torn from the fences was whirled high in the air. Sheaves as we handed them up were nearly snatched from our grasp and hurled far to leeward; heavy gusts struck the hand-rick, now packed and solid, making it shake and stagger; and Peter, perched on the topmost step, flung his arms about the narrowing point to keep it fast while we ran round the base with ropes, fixing it down in ever-tightening spirals. We made two of these hand-ricks in the morning; after dinner the wind moderated and we built several more of various shapes and sizes. Three small ones, which had been made out of the corn cut on the first day, were found to be damp, and we took them into the sawmill to be dried on the now vacant hay rack. This job was hardly done before darkness overtook us, and we were forced to leave a fair number of fallen stooks to be set up again on Sunday. As anticipated, it poured on Saturday night, but our strenuous building of hand-ricks was rewarded by the sight of them standing undamaged by wind or rain till the end of the month. The rest of the corn in the long cornfield was gathered into large hand-ricks by 14 September; the last of these was carried in good condition on 2 October.

We enjoyed this struggle with the weather. In its constant demand for watchfulness, resource, and sheer hard work our primitive harvesting yielded some part of that grim satisfaction and delight in sporting hazard which is felt by the sailor in his dealings with the sea. Nor was that treacherous and seductive element ever very far from our thoughts; its power could be felt as a subtle influence rising and falling in the recesses of our consciousness, just as its stealthy tides would ebb and flow in the landlocked waters of the Highland lochs. On the farther side of Skye lay the open ocean, and the low spreading trough between the Cuillins and the hills of Sleat let through Atlantic gales to wreck our stooks and dry them too, for I am convinced that without that everlasting wind we should not have saved a quarter of our crop. Perhaps from the purely financial point of view we need not have troubled ourselves so much. Not many pounds were at stake in those two small fields, and we could have bought enough cake to have taken the Belties through the winter, as they could get plenty of roughage in the shape of heather and coarse tussock grass. But somehow our credit was at stake; we meant to beat the natives, and also to show the Laird that, even if we had sown too much, we were able to secure it without great loss. Besides, corn production of any kind seems to have a special fascination for all of us, so that to the layman it is agriculture *par excellence*. Its technical terms, like those of seamanship, have become so deeply embedded in our common speech that the actual meaning has been lost in the metaphorical; and for those who read the Bible there is a wealth of poetic imagery that comes direct from the cornfield. Bread is no longer the staff of life; in most families it is a dull and despised medium for the consumption of something else. A person is 'bread-hungry' only when he cannot get anything better, and the baker's bill is usually the smallest item in the housewife's weekly budget. Yet our devotion to corn is so great that we subsidise arable farming in a country which is far better fitted for the production of livestock on grass.

And if Britain as a whole is not a corn-growing country, still less so the Western Highlands. Wheat is of course impossible, and even oats are speculative, and should never be grown by anyone who does not enjoy a sporting risk. If you want safety, stick to potatoes of an orthodox variety like Kerr's Pink, from which a fair crop may be expected in almost any season. In the old days, when failure meant famine, there must have been terrible anxiety. But now, when modern transport and potatoes have made the

Highlands independent of home-grown grain for food, we can watch the soaking oats turn black with nothing more than the distress that any farmer, however well-provided with other things, must feel at the loss of a crop he has worked for, however unimportant. Most Highland farmers grow corn only to supplement bought feeding-stuffs or to make a rotation with potatoes; it is useful for feeding to stock, but too precarious to be worth cultivating for that alone. It is curious to reflect that, having ceased to depend on their own grain for food, they can raise better and larger crops than their ancestors, whose need was so much greater. In the eighteenth century, the West Highlanders cultivated their corn with a wooden spade, harrowed it with a rake tied to a horse's tail, or more often, drawn by a woman; and as they never kept any but the worst grain for seed, they would often get no more than threefold or fourfold returns.

Yet, though we no longer eat the grain we grow, corn is the very essence of harvest. The sentiment lingers long after the reality has gone. No food for beasts alone can stir so vital an interest as the potential food of man. From the day of its sowing the corn-field absorbs my attention. I love to watch the first delicate shoots, grasslike yet totally different from grass, brilliantly green against the dark wet soil; to feel the brush of the strengthening stems against my legs as I move through it with a scythe, lopping off the heads of docks and thistles; to see its rippling movement when the wind passes over it, for corn stirred by the breeze is more like water, and blown grass as the hair of a long-coated animal. Then comes the deepening green where the crop is extra heavy through uneven manuring. After a stormy night, when rain-squalls howl in the chimney, I look out with anxiety, expecting to see the heavy parts laid flat, as if a giant fist had crashed down upon them from above. Later a yellow tinge creeps over the field; certain parts on the edge show very white, for the birds, emerging from their cover, have cleverly extracted the grain. The White Line straw is so stiff that it will support the fattest chaffinch perched upon the ear, and sway gently with his movements as a foxglove sways when a bumblebee rifles its bell: and so the birds need not wait for their feast till the corn is in the stook.

The second period of our harvest began at noon on the 10th, when we started to cut the field below Rattray's. This, though sown a week later, had overtaken the long cornfield, so that we had to work the two fields concurrently, the earlier having reached the stook and hand-rick stage when the latter was still being reaped and bound. The land, which had been well harrowed and rolled in the spring, was smooth and clean, and except at the western end, the corn was not too long in the straw. There were no stones to blunt the scythes and no thistles to delay the binders; we made fine progress until the usual afternoon deluge stopped all outdoor work for the day, and Vivian was dispatched to hunt for food in the village. Next morning we resumed cutting, but it began to rain soon after dinner, and we retired to the sawmill to fork hay. The previous year we had the corn in stacks outside, but the prolonged bad weather made me wish to have as much as possible stored under cover. We therefore decided to shift all hay from the southern end of the sawmill to the northern, thus leaving a large space that could be packed to the rafters with sheaves, and would if need be hold the whole crop. This decision was most fortunate, for at no time during the following six weeks could we have built three or four stacks with safety, quite apart from thatching them. And the sawmill not only sheltered the corn when stacked, but provided a rack on which damp or doubtful sheaves could be dried out, thus lessening the risk of heating. The 12th of September was a fine day, but with the exasperating casualness of casual labour, Murdo and the shepherd's nephew failed to appear, and being two scythes short, we went to the long cornfield and made hand-ricks.

The 13th was the finest working day of the season, and we made good use of it. There was no wind, and the corn was drenched with dew, but after so many delays we could not afford to wait for ideal conditions. So I ordered an early start, and the first sheaves were cut in a state of moisture that would have appalled a southern farmer. At the beginning of the harvest I should not have done it myself, but we were getting a little desperate, and beyond tying the sheaves loosely and stooking them in fours we left the thing to providence. The two hired lads went at five to help a neighbour cart his hay, but the rest of us worked late, and when we finished there was not much more than a day's work left to do. It was a strenuous business. The field was in a hollow, surrounded by woods; not a breath stirred, and the sun beat down on our backs as we stooped to bind. Fine days were rare, and we had to hurry. At four o'clock weariness seized me, and for the last two hours I worked in a daze, lashing myself on with the thought that at the end of so many swathes there would be rest and tea. This day was a test of strength to those unused to outdoor labour. Mr Gordon was completely done, and retired to an early bed, while Vivian bound his sheaves slower and ever slower, and began to lose interest in harvesting. Small blame to him or anyone else! After tea Peter horrified Herself by walking the eight miles to and from Strathascaig to discuss sheep farming with the Laird.

The next day was hot and oppressive. Once more we started cutting early, but soon after eleven thundery clouds were seen in the west, and just before twelve a distant peal was heard. We made a rush to the long cornfield and secured the rest of the stooks in hand-ricks. Among the hills of Skye a storm was in progress, which at any moment might drift in our direction; but after causing a commotion of curses and hasty work it passed away to the north, and the rest of the day was dry.

The field below Rattray's was finished next morning, and the last sheaf carried home to Flora in the kitchen, where it was kept until defaced by mice. Murdo immediately reset his scythe and went off to cut more hay in the sawmill park, while the shepherd's nephew, assisted by Vivian, fitted Dick with a set of ready-made shoes. I had a suspicion that some of the hand-ricks in the long cornfield were damp, and went out with the intention of opening one or two of them and spreading the sheaves in the sun. I had not taken the rope off the first when I heard what sounded like a distant roll of thunder. It came from the east, the quarter from which thunder most frequently works up. The view in this direction was obscured by trees and nearer hills, but there were darkish clouds above them which might well be the vanguard of an approaching storm. It had thundered yesterday, and we had seen summer lightning at night. I did not risk opening the hand-ricks. Several more peals were heard, all in the same direction, but no storm came. Next day I learnt that there had been naval manoeuvres at Invergordon! Peter, who was a great reader of newspapers, declared that it served me right for stopping *The Times* in summer; but why spend threepence-halfpenny a day on a weighty publication that no one has (or should have) time to read, printed on paper which is notoriously unsuitable for lighting the kitchen fire?

From this time onwards small loads of corn began to trickle in to the sawmill. The first couple of hundred sheaves were from lack of experience stacked too damp; they heated and we had to cart them out to the fence to dry. After that we made a practice of feeling each individual sheaf as it came in; if any one showed a trace of dampness, it was put on the rack with band loosened, or in bad cases completely untied. We had cherished a faint hope that those three fine days would be the first of a spell of settled weather. But the rain came back to stay, and by the 24th it was clear that unless something were done promptly the corn in Rattray's field would sprout in the stook. The field was too sheltered for rapid drying; the stooks were sodden and drooping, grass was growing through the butt-ends, and many sheaves were already sprouting under the bands, though the heads, except where stripped by pigeons, were still in good condition. I regretted now that I had not carted off the whole crop as soon as it was cut and stooked it in the long cornfield, where there was always plenty of wind to dry the stooks, and bare stubble under them, instead of the lush undergrowth of sown grass that spoilt the drying in the other field. The byre drain had choked, and an old man had been sent by the estate to dig a trench for new pipes. Before settling on a croft in the village, he had gained years of experience as a farm servant on the east coast, and I thought he might have some useful suggestions. I took him up to look at the field. He agreed that the corn could not be left where it was, and advised us to cart it to the sawmill field and let the sheaves dry upon a 'soo'. This is an old device, well known in the north as an effective substitute for stooking in bad weather. It is a steeply pitched trestle-shaped erection, roughly built of old posts or odd pieces of wood lashed or nailed together; it can be extended to any length desired, and raised to the height at which a man can conveniently reach to fill it. The sheaves are put on the soo in overlapping layers, heads inward, in the manner of thatching, while the ends are left open for the air to pass through and dry the ears. The topmost layer of sheaves are inverted and tightly bound with a rope. We had a quantity of spars left over from the tripods; they were hauled out by Peter and Mr Gordon, who rapidly built them into two soos which held between them about 1,600 sheaves. Meantime Vivian and Murdo carted down the corn and I put it on the soo. But this was not the end of our trouble. To save marking or splitting the spars, which were new and squared, we had on the old fellow's advice used lashings of stack-rope instead of nails. The open ends of the first soo had been strengthened with wooden stays, but the weight of the sodden corn was enormous, and before long ominous groans and cracks were heard, the middle of the soo began to sag, and presently the whole erection quietly collapsed, in torrents of rain. It was soon rebuilt with nails and internal stays, and the second one to match; after that they weathered a gale and we had no further misadventure. But I do not think that we should build soos again. It is true that the sheaves, though wringing wet when put on, did not heat, and only the topmost roofing layer sprouted. But the drying was very slow, as, apart from the ears, which were exposed to the wind passing through the tunnel, there was too little ventilation of individual sheaves. The weather was always too bad to empty the whole soo and spread its contents, so that we had to clear it piecemeal at the rate of twenty to fifty sheaves a day, most of which needed further drying on the sawmill rack before stacking. The soos were built on 24 September and finally cleared on 12 October. We lost a certain number of sheaves from the top, but the majority were secured in moderately good order.

The two soos when full would not accommodate more than half the corn from the field below Rattray's. On the 27th we decided to try the experiment of drying the remainder on the fence like hay. The south side of the long cornfield was bounded by a fence of two wires only, the lower one raised well above the ground. To this fence the sheaves were carted and put on vertically, the butts resting in serried rows upon the lower wire, which split them to the band, while the heads were tucked firmly under the upper one. Thus jammed they were secure from storms, and being unprotected and broadside on to the prevailing wind, they dried far more quickly than their neighbours on the soo, though need-less to say they suffered several fresh wettings before we could get them secured in hand-ricks on October 2. All this corn was wringing wet when it came out of the stook, and most of it was put on the fence in a pitiless downpour, but it was finally saved in better condition than any other. If at the time of cutting I had been able to foresee the appalling weather that was to come, I should have put every sheaf on the fence without stooking at all, and allowed it to mature there; for I have no doubt that this method would have saved much loss of time and labour.

By the third week in September we realised that we must be ready to finish drying all the corn indoors. As we were still making hay, the barn with its rack and packing accommodation was reserved for that, leaving the sawmill free for corn. The big sawmill rack was too small for our needs, and we were busy contriving new drying-places for the sheaves that came in daily from fences, soos, or hand-ricks. We made a small rack in the stable, but the sheaves dried too slowly, and it was abandoned. The granary, a long upstairs building covering the whole length of cart-shed and dairy, was converted to corn-drying, a small space on one side being reserved for Mr Gordon and his poultry food. The window, which was in line with the door, was taken out, and light racks of sheep netting stretched across the rafters. There was an excellent through draught, and sheaves could be unloaded into the cart-shed below and forked straight up through a trap door. The cart-shed itself was fitted with overhead racks of wire and sheep netting, while surplus sheaves were spread about on the wood heap, draped over the trap, or leant in a single row against the wall. Thin planks were laid across the rafters of the sawmill, with a space between each, and a layer of sheaves one deep was spread across them. The whole farm seemed buried in half-dry corn – every place was littered with odd bits of straw, husks, and shaken grain. The poultry were in clover; not only did they get these droppings, but any sprouted or mouldy sheaves were taken out for them to scratch over at leisure. I began to think that we should soon be drying corn in front of the fire or in our beds, so firmly were we resolved to win this battle.

Peter and I would spend one or two hours daily in testing, sorting, and grading sheaves into three classes – those ready for stacking, those needing a spell on the rack, and those only fit for poultry. The second was the largest class, and the third fortunately very small – perhaps one cart-load out of twenty. Although the last load came in from the field on 13 October, the sorting of sheaves was not finished till four days later. A large number of sheaves came from hand-ricks, and many of these were too damp to stack, simply because we were hardly ever able to open and spread the hand-ricks before carting them indoors. And even if we succeeded in spreading them, it would be only for a precarious hour or two between heavy showers, at whose threatened approach someone would rush out with a tarpaulin and bundle the sheaves under it. On many days the mere taking in of a hand-rick without previous spreading was impossible without a cover on the cart, and even then much speed and calculation were required to dodge the frequent and torrential downpours. At the first easing of the deluge, Peter and Murdo would drive out to the field, stand by with the cart till the rain ceased, fork up a hand-rick at furious speed, pull on the tarpaulin, and return to the steading. Many a time another shower would start before they could fork off the load, and the cart must stand loaded and covered at the sawmill door until the next interval. Nearly all the hand-ricks had been built against time, and were too leaning or flat-topped to run off the water. In many cases the top layer of sheaves had become black and mouldy, and these, together with the upper section of the soo, made up the 5 per cent of lost or condemned corn. The best of the hand-ricks, which had been made during the three fine days in the middle of September, were used to build our only stack outside, but of this more later. We had our hands full without stacks. Sitting in the byre at our cows, Peter and I would sing to the accompaniment of rain drumming on the iron roof, a new version of the well-known harvest hymn:

> Black waved the mouldering corn
>   In Ross-shire's rain-drenched land,
> When full of 'damns' one streaming morn
>   Went forth the reaper band.

But we saved our crop; and now that the Belties munch the rustling sheaves in the mellow winter sunshine I am glad we worked so hard.

17

October: The End of the Struggle

DURING all this time our existence was very self-contained. We never left the farm except to go for Sunday supper to Strathascaig, and this could hardly be called a social function. Flora went twice a week to her home in the village and brought back the latest gossip. Mr Gordon walked to various places in search of provisions, and would often have some news to give us. The shepherd came down to meet the post, and there would be an informal gathering in the cart-shed, or people with cows for the bull stayed to have a talk with Peter. On rare occasions carriage company came to call, and there would be a stampede to tidy ourselves and clear some place to show them into. Peter, with the selfishness so common in such emergencies, would hide in the barn and abandon me to deal with the invasion alone. This was annoying, since a life as quiet and yet as busy as ours leaves the wheels of polite conversation a little rusty. I could rarely think of anything much to say; it was not shyness nor even boredom, but a feeling of being somehow in the wrong box, with bare legs and no cake in the house. Sometimes it would be the landlord, who asked a few questions about the cattle, ordered a pair of table cockerels, and had the sense to go away quickly.

I am afraid that all this sounds rather misanthropic, and must confess with shame that neither Peter nor I are gregarious, which is perhaps one reason why we find the country so completely satisfying, even without a car, or wireless, or weekend visitors. We love a talk with a friend, but formal company and society contacts fill us with apprehension and weariness of spirit. Social intercourse, if there is to be any vitality and pleasure in it, must be spontaneous and personal; and it is the want of this essential merit that makes the conversation of the average well-bred person so lacking in human interest. In this country we are trained never to give ourselves away, a faculty which may possibly make us discreet officials, but will certainly render us dull companions. The man who never gives himself away must always be on his guard against his own feelings, and this studied reserve will lead to excessive self-consciousness, frigidity, and convention in thought and speech – in fact to all those unnatural and unnecessary inhibitions that freeze the springs of good fellowship at the source.

This excessive fear of giving himself away will often cause an educated person to avoid like poison anything that smacks of the personal. He has a secret longing to talk in the first person singular, but feels obliged to keep to the third; he prefaces most statements with a careful 'apparently', and prefers the vague 'it seems that' to the plain intimacy of 'I know'. No one is really deceived by these phrases; the personal opinion, however discreetly veiled, is still there, waiting for applause or condemnation. He gains nothing by his evasions – he merely loses strength and authority, and worse still, becomes dull. Compare two conversations in a railway carriage, one with a well-bred, well-educated member of the upper classes, the other with a working man or woman with no pretensions to breeding or education. The first will talk of the weather, the scenery, the news, and at the end of it all remains an abstraction, whose real nature is completely unknown. He has talked for two hours without giving away one particle of his essential being. He has lost no dignity, but neither has he gained a friend. The second talks of himself and his work, his children's ailments, his mother-in-law's temper, his dog, his cow, what he said to her and what she said to him; and at the end of it we have come to know not a figurehead but a man, with all the funny, intimate, miscellaneous details that go to the making of human life. There can be no doubt as to which of those two conversations contains the most interest. Even historians have given up their preoccupation with dynasties, campaigns, and policies, foreign or domestic, and allow us to enquire how the Anglo-Saxons ploughed their land and how the Elizabethan sailor handled his ship.

That is why I shall never apologise for making this book a string of personalities. To write an impersonal book is an impossibility, and if it were possible it would be duller than the multiplication table. The driest treatise on the differential calculus, by John Jones, is only apparently less personal than the confessions of Daisy Darling the cinema star. The treatise might really be labelled 'All that Jones knows or thinks that he knows about the differential calculus'. It is the fruit of years of labour and controversy, and is packed with emotion, however carefully concealed. Moreover, we can only get at the subject as presented by and through Jones' mind, and further distorted by our apprehension or perhaps misapprehension of what he is trying to tell us. As this train of thought, if logically pursued to the end, would lead to complete scepticism and practical lunacy, I shall leave it alone, having used it only to demonstrate the impossibility for us as individual persons to escape from the personal.

The society of the Highlands has little of the professional or middle-class element. At the top are the 'county' people with their shooting-lodges, Highland gatherings, and discreet patronage of local charities and industries; at the bottom the crofters, shepherds, and villagers. Between these is a void, in which float in rather nebulous and chilly isolation a few doctors, ministers, bank agents, and large farmers. These of course have their relations upwards and downwards, for Scotland is a democratic country, but they are not numerous enough to form a compact and vital society of their own. The gentry of the shooting lodges, taken as a class, are not interesting. I do not see much of them, partly because I have neither money, clothes, nor car, and partly because the Highland season corresponds with the busiest period of the farmer's year, when I could not, if I would, find any leisure for society. But what I have seen does not make me want to probe further. I am not sure if this feeling is unconscious Bolshevism, intellectual snobbery, or mere sour grapes. Perhaps I dislike seeing them roll by in cars while we are wrestling with the hay, or having to listen politely to drawling inanities when overwhelmed with urgent jobs; or it may be that I am conscious of being regarded as at the best eccentric, but more probably a borderline case or a social black sheep, and resent this libel on a hard-working farmer, whose only crime is lack of money, quiet tastes, and a turn for practical agriculture. Compared with the production of young pedagogues, calf-rearing must seem (to borrow one of Colin's favourite phrases) rather 'low' – a kind of apostasy or betrayal. A few months ago, when revisiting the university at which I once taught, my neighbour at the lunch table, whom I had known previously, after a longish pause in which she was obviously wondering what on earth to say to me, remarked drearily, 'I suppose you still keep cows.' I admitted that I did, whereupon she said 'Really,' and turned to the person on the other side. Thus is the middle-class farmer who does his own work a kind of Cinderella. But it does not matter; the more often the cars of the leisured flash by without stopping at the farm, the less time will be wasted on polite chatter that leads nowhere.

And now back to work. October came in. At last, we thought, the gods will relent and send us a spell of quiet air and mellow sunshine, in which we can refit for the last stage of the campaign. The days will be short, but long enough for our weary bodies, and if the rain will only hold off for a little we can still make up for lost time. The first two days seemed to justify our hope, but after that the weather became rapidly worse, and the last fort-night of October beat all records, combining summer gales and downpours with increasing cold and darkness, hail, sleet, and even deep snow! But the reader must not imagine that these conditions were anything but abnormal. Even in my compara-tively short experience I have seen *single* harvest months as wet and stormy as those in 1934; but never in living memory have there been fifteen whole weeks of wind and rain without a break worth mentioning – certainly not at this time of year. If such weather were the rule, no one would farm at all. The drought in England still persisted, and while we paddled round our sodden stooks, English farmers, who had secured a quick and early harvest, were bawling for rain and more rain. My mother, writing from Essex, complained that everything was burnt up, and the corn had been carried so early that the fields looked bare and uninteresting. Would to heaven that our fields looked bare and uninteresting too! There is no fairer sight than ripe corn rippling and swaying in the wind, or a field in the stook dreaming under the moon. But danger lurks at the heart of that beauty, and the farmer longs for the clean and naked lines of empty stubble, and the peace of accomplished purpose.

With the idea of saving space indoors, the Laird advised us if possible to build one stack outside. Peter was anxious to try his hand at stack-building, and as it was difficult to secure an expert, and the only one available was tediously slow and expensive, we decided to do the job ourselves. Murdo had often seen stacks built, and was consumed with longing to try himself, but he was afraid that the older men would come up and criticise his handiwork, and I persuaded him to make the attempt only by promising that the stack should be taken into the sawmill when the rest of the corn was dry, and need not stand all winter.

The best local stack-builder lived in Murdo's own township. I had known him for a long time, as he always stacked the Laird's corn and in 1933 had worked for me also. He made a neat job of it, but was lamentably slow – taking two days to build a couple of ten-foot stacks, and two more to thatch them, though he had Murdo and Alec to cart the sheaves, and one helper to hand them. In all this land of blether, he was the supreme bletherer; and that endless and often incoherent eloquence, combined with a venerable white beard, had earned him the name of the Prophet. Early hours never suited him. When farming on his own he was said rarely to rise before twelve, nor did he care for any unusual exertion. One day, when driving on his daily milk round, he found a tree blown down across the farm road. On the sodden grass verge, which was periodically swamped by an over-flow from the ditch, was just sufficient space for the cart to pass. Sooner than saw the trunk across and move it, he drove round, day after day: I myself have seen the tracks. Had it not been for his excellent and energetic wife he would have been in a sorry plight. Small wonder if he was a little henpecked, and, like a shepherd's well-trained collie, kept to his own corner. 'It is easy to see,' said Murdo to Alec in their famous discussion on marriage, 'who is the boss in that outfit.' He had sent a calf to Dingwall, and I asked him how much it fetched. 'It was herself sent it,' he replied, 'and she didn't tell me the price!' This may well have been a tactful way of evading a blunt question, but no cave man would have made so damaging an admission, even if it was not true – especially to a woman.

My earliest memories of stack-building, in the happy irrespon-sible harvest work at Strathascaig, when the crop was small and of little importance, are bound up with the Prophet and his neat, finicking activities. When a good day came and the corn was mature, we hunted round the steading to find poles of which to build the central tripod. Most of them were too long, too short, too rotten, or too deeply committed to some other purpose. At last three were collected, one, originally intended for an aerial and not required, two, a discarded part of the hayframe, and three, a nice piece of ash cut in provident mood the previous winter to make handles for derelict picks and hammers. The tripod, like some huge camera-stand, was set up, its points lashed together with rope. Lateral bars were nailed on, and a tunnel made at the side to let air into the centre. Round this tripod the Prophet would build the stack.

Meanwhile Colin and I went off with the horse to gather the corn. I stood up in the cart, while Colin tore the headbands off the stooks and forked up the sheaves almost faster than I could arrange them. Hurtling masses of corn struck me; I was buried under mountains of bristling straw. Trampling it underfoot, ranging it symmetrically on the frame, I subdued it, and finally emerged on top, precariously balanced. Colin led the horse, whose ears alone were visible from the summit of the load. The cart groaned and swayed across the uneven stubble. It was drawn close to the tripod: Colin handed me the fork, and I in turn hurled sheaves towards the Prophet, aiming them head to centre, a little in front of him. Being a woman, my aim was sometimes rather wild, and the sheaf hit his white venerable head. But he did not seem to mind. He arranged them with all heads pointing inwards. The stack got higher; we emptied the cart and went for more. The Prophet's level rose until, whereas I had once been forking down to him, I was now forking up. He began to decrease towards the top. Now he crouched on the sloping side like a slater on a roof, and grabbed the sheaves from the point of the fork. Finally he crowned the stack with a bundle of corn, a kind of super-stook; and, perched aloft, surveyed the world like a man at the mast-head. My elevation in the cart seemed very low. Slowly and carefully he put the finishing touches to his handiwork. A ladder was brought for him to descend. Taking a rake, he combed down the sides of the stack with lingering care, as a mother combs a child's golden hair. 'It will do,' he said, with conscious pride. Colin watched him with admiration. He too was of the slow, careful type, and the Prophet's delays did not annoy him. Moreover the Venerable had just taught him the four and five plait in straw with which he used to decorate his horses' tails when, long ago, he won a ploughing match.

I was tired, and in dreamy mood watched the sun slanting across the bare stubble, and the tide, now close upon the equi-noctial springs, brimming at the foot of the dyke. High in the air a harsh note was heard: a pair of ravens, mere specks against the blue, passed overhead. The wild was all around us; the rocks in which they bred were unchanged from the beginning, when every harvest field was pathless moss. A flock of crows settled on the stubble. 'Uncle!' Colin exclaimed, his mind released from work like a stone from a catapult. 'Can I get the gun and have a shot?' 'If you like,' replied the Laird, and giving a hitch to his kilt he walked towards the house, contemplating the delights of dinner and, after that, a roomy and slumbrous armchair before a fire of green birch blocks.

For the next few days we kept thrusting long arms into the stacks to see if they were heating. And sure enough, one of them was a little warm and yeasty inside. It is said that when things go wrong on a ship it always happens in the second mate's watch. And this rule sometimes applies on land. Those in command have a habit of suddenly and mysteriously disappearing from the scene of action, either because the post is seen approaching, or somebody's car is heard on the road, or to whistle in a dog, or to get a drink, or for some other reason best known to themselves; and during one of these inexplicable absences, when I was left in charge of the field, the Prophet chose to make one stack where he should have made two; and I, supposing him to have received special orders, said nothing. This over-grown stack was now a trifle warm; with many a curse we pulled it to pieces, left the sheaves awhile to dry, and then rebuilt. After this, all went well. A few days later we cut the rushes that were providentially growing on the farther side of the field, bundled them roughly and carried them over to the stacks, which we had fenced round with barbed wire to keep off the cattle. They stood in a yearning circle, and Pendeen, that ardent searcher for new and exquisite food, thrust her giraffe-like neck through the wires, and shot out a sinuous questing tongue, if by any chance she could draw in a juicy corn stalk lying just out of reach.

A ladder was brought, and the Prophet started to thatch the stack with rushes. He took a good handful, pulled them level and thrust them well under the straw, letting each row of rushes overlap the one beneath. Meantime I wound the stack rope into large balls. The Prophet stood on the top of the ladder; I walked from side to side, throwing him the ball of rope; he passed it over, while the Laird cut the ends and made them fast. This went on until the top of the stack was covered with a coarse network of rope drawn tight across the thatch, fit to withstand the worst gales of winter.

The stack-building at Achnabo was done more quickly and under far less pleasant conditions. The 2nd of October was a fine day, with no suggestion of permanence – a 'pet day', as the saying is, and we had to hurry. Peter and Murdo carted the driest of the hand-ricks from the long cornfield and made the stack together, Murdo building and Peter serving him with sheaves. The rest of us were busy putting the sheaves from the long fence into hand-ricks. Murdo's stack was quite good for a first attempt, but the head was too elongated, and he had failed to keep the butts of the sheaves sufficiently pressed down. The chief fault of the stack was a mere concession to neatness, learnt no doubt from the Prophet. The butts of the sheaves were very ragged, and in order to give the stack an even and tidy finish, Murdo had turned in all the straggling ends, so that the water, instead of being thrown off to the ground, was carefully conducted along the doubled straws into the heart of the stack. I had an idea that this was wrong, and nearly stopped it. I wish I had, for when the Laird came to inspect our handiwork, he fulminated for several minutes and Murdo was told to pull out all the ends again. Rushes were cut, but the stack had taken the water so badly that it could not be thatched. Day after day we waited, storm upon storm beat down and soaked its naked, blackening head, now furred with a vigorous growth of sprouts, insolently green and brilliant. On the 22nd it narrowly escaped destruction, and on the 26th Peter began to take it in piecemeal with the help of the tarpaulin. Much of the top was rotten, mildewed, or sprouting, and had to be given to the hens, but the lower part was in good condition. The whole stack was cleared indoors by 7 November, five weeks after building, so that Murdo had his wish: it was never surveyed by carping experts in the leisure hours of winter.

I came back from my holiday on 4 November to find that Peter and Mr Gordon had started to lift the potatoes. There was no elaborate ceremonial; we had no outside labour, and did not even use the plough. We took a drill or two at a time, as the weather allowed, lifting them with a graip and carrying the bags to the barn on our shoulders. The ground was sodden and chilled with melting snow, the tubers thick with mud and cold to handle, so that Mr Gordon came out to work with gloves and a little trowel. It was, however, neither the latest nor the coldest potato-lifting I have known. One year at Strathascaig we finished on 16 November, when the sun had left the field for its sixty-days dip below the southern hills. The northern side of the glen was bathed in sunshine, but where we worked it was freezing all day, and our hands and feet were like blocks of ice.

The Laird grew a fair quantity of potatoes. He supplied various neighbours, and the shepherd must have had nearly two tons as his perquisite, many of which were thrown away, for Highlanders are as wasteful with potatoes as English housewives are with bread: except in time of war, a staple food seems to have little value. They were lifted with the plough, and a large number of helpers was assembled to gather them into pails and fill the bags. By this time Colin, to his great relief, was back at school; but the shepherd was there with his son, and whatever maid was in the house, and Willie the Scout, and Donald the crofter, and myself, and intermittently the Laird, and if it was a Saturday, two school-boys from the village, and always a strange pair of spinsters from the township on the shore, Annabella and the Slug. Annabella was supposed to be the dirtiest woman in the parish; no one liked to sit near her at a fire, and she was said to clean the byre with her hands. She would herd her beasts on the moss, a weird figure in strange flapping clothes, attended by a dog as wild and battered-looking as herself. The Laird had always had a weakness for the less elegant and respectable members of society, and his love of tramps was a thorn in the flesh of his wife, who liked order, decency, and fixed habits. His last resident handyman had been a tramp who came demanding a job of work and stayed for the rest of his life. No one asking for food or shelter was ever turned away. I remember Mr William Paterson, a lordly gentleman of the road, who used to visit Strathascaig twice a year and sleep in the stable, sitting on a tea box with his head sunk between his knees, for he never lay down. When I went last thing to feed the horse, and switched on the light, he raised his head and told me crossly to put out the light and see that they had his brose ready at the door in the morning. Another crony of the Laird was a man who lived all winter in a cave by the seashore and went about the district sharpening and setting saws. An interesting fellow; but he died the other day, and the last saw-sharpener who came our way spoke with an English accent. He had lost his job in a Sheffield cutlery works and had two daughters who made their living by sharpening butchers' knives.

There was nothing of the vagabond about the Slug. She was a very large, heavy woman, not so much fat as massive, and at the same time soft – hence the name. Soft in nature as well as in body, having never left home or been obliged to stand on her own feet. She was now in her fifties, and still kept the cottage her parents had lived in, with a few hens and a patch of potatoes, supported by remittances from a brother and the proceeds of a little casual work in the neighbourhood – plucking, scrubbing, knitting, and the manufacture of white and black puddings. For want of anyone better, my mother once employed her as a daily servant, but her slow, lumbering methods, or rather lack of them, drove us to distraction. Our English accent baffled her, and she was either too proud or too polite to admit it, so that we were never sure whether she had understood or not, until being told to do one thing she promptly did the opposite. We should, in sailor fashion, have made her repeat her orders, but this I am sure would have given great offence. Poor Slug! she was a good soul, and rarely came without some little offering, and – strange merit in the West – was never late. Herself employed her for a weekly scrub, and was exasperated by her lack of dash and muscle; but she was in far too soft condition for vigorous work, and was always burdened, summer and winter alike, with innumerable layers of vests, petti-coats, skirts, jumpers, cardigans, overalls, and shawls, which only fostered her lethargic habit of body and tendency to chronic colds in the head. Nor did all these wrappings keep her warm, for no matter what the season, she was always complaining of the cold, and questing vaguely round for a cup of tea. Three times a week she called for her pint of milk in a whisky bottle, and on these occasions she would wander into the kitchen, and as she sipped her tea would watch the labour of others and comment on the coldness of the weather. To which Herself always longed to retort that if only the Slug would work harder and faster she might feel warmer. But the influence of half a century of crawling cannot be undone in a day.

The Slug's supreme annual effort was the lifting of the Strathascaig potatoes. No doubt she filled far fewer pails than anyone else, but at least she was present all day, and fearful was the backache of which she complained in the evening. After that consummation she, and we also, could retire to our long and peaceful hibernation. We always tried to get the lifting done in one day, which was quite simple if there were plenty of workers and no rain, and if the Laird remained in the field himself the whole time. It is true that he did not do so very much – filled a few pails, tied the bags, kicked about in the drills and gleaned a few overlooked tubers, watched the sky, consulted with the ploughman, whistled his dogs, and above all exercised that mysterious but potent influence for good we call the Master's Eye. The holy man who first hung in a dark corner the text 'Thou God seest me' was a profound psychologist. If we were sure that no master, human or divine, was ever watching us at work, there would be very little done – very little indeed.

The mail train, by which visitors most frequently arrived, reached the station at 1.15 p.m., and to meet it the Laird was forced to leave the potato field at the busiest time of the day. When the decreasing hum of the car died away into the distance, Donald, having opened a drill, leant against the near horse, lit his pipe and began to think. The Slug collapsed upon a half-filled bag and drifted into vacancy. Two urchins from the village started pelting one another with potatoes. The shepherd straightened his back, and scanned the horizon for some sheep in need of instant attention. But none were in sight. Glancing at his own cottage, he noticed a thick plume of smoke ascending from the chimney; his elder son, returned early from work, had probably put on a few fresh sticks to boil his tea kettle. 'I am thinking the hoose will be on fire,' he remarked to the company, and made off, followed by his youngest. No one seemed interested in the problematical fire; the Slug dozed on her bag, the boys continued to pelt one another, and Donald resumed his meditation beside the idle plough. Willie and I were left alone at work. 'Dash it!' said Willie, who disliked the shepherd, chiefly because he never did his share when double-sawing; and then in his indignation forgetting the presence of a lady, 'the lazy b—— !' Presently the train appeared, and in a few minutes we heard the sound of the returning car. The shepherd came back from his fireman's duties, Donald hitched the horses round and started opening a fresh drill, the Slug rose reluctantly from her bag, and once more potatoes were thudding and pattering into the pails. A half-ton load was ready to go in; one of the plough horses was harnessed, bags were piled on the cart, and off it went, grinding and jolting over the rough road, through the ford and up the steading to the barn, where the bags were emptied on the floor to dry. And then back to the field – more drills lifted, pails emptied, bags filled, tied, loaded, and carted home. And then to sleep.

The smallness of the crop, the absence of outside workers, and the piecemeal fashion in which it was performed, made our own potato-lifting seem something of an anticlimax. In theory it should have been the crowning mercy, the grand Amen of the harvest, but in actual fact the last few drills were lifted when we were still taking in Murdo's stack and carrying a sledge-load or two of stale hay, so that the emptying of the final bag onto the barn floor passed uncelebrated and indeed almost unnoticed. The snowstorm of 31 October delayed all work, and when I returned from my holiday on 4 November, I saw a few dismal coils of hay still out, with snowy tops and a look of shop-soiled Christmas cakes, a half-dismantled stack protected by a tarpaulin, and five or six drills of unlifted potatoes. In the hayfield was Murdo, who, while waiting for me to stamp a long overdue insurance card, was trundling a huge snowball among the coils. But it was already thawing, and, helped by rapidly improving weather, we gathered in the tail-end of our various crops, trimmed the packs of hay and corn, covered the potatoes, hauled up the racks, gave the rushes intended for thatching to the bull for litter, and settled down to enjoy our winter. It was finished. Thank God.

Our pleasure was lessened by the poor results of the autumn sales. The temporary fillip given to prices by the announcement of the beef subsidy had come to an end; there were too many stores in the market, no one knew what would happen about the Irish cattle, and the failure of the turnip crop had made farmers unwilling to buy animals for fattening. An excellent weaned calf made £6: 10s, an eighteen-months bullock £4: 15s, and the two barren cows £3 and £1 apiece. I sold a cross-bred heifer calf three months old to a local man for five shillings less than the bullock had fetched in the open market; but the saving of the harvest made it possible to winter the other nine calves on the chance of an improvement in the spring. Since then, agriculture has once more been sacrificed to industry, and the coming invasion of Irish cattle will not help us to make good our losses.

My holiday was mostly spent in the trim rural beauty of Gloucestershire; apart from a few welcome showers, the weather was quiet and fine, and Nature played in so subdued a key that the furies of Achnabo seemed a fantastic dream. Into this haven of peace came Peter's letter describing the famous storm of 22 October. It was read to a group of incredulous friends. Vivian had sent a bundle of his photographs, in which we were displayed scything and filling fences, and there was one which showed the long cornfield levelled by the gale of 8 September. They were handed round and inspected with mingled wonder and amusement. What could I possibly find in such a life? And dash it, as

Willie would say – dash it – I don't know how to tell them without appearing a hypocrite or an idiot. When I first went south I was tired, and thought how pleasant it would be to live in a place like that, where cattle browsed peacefully behind well-kept hedges, where hay could be made on the ground and corn dried in the stook, where feet are often dry, and a good town five miles off. Yes, it would be pleasant. And when we left Achnabo, perhaps . . . But those were the thoughts of a tired woman. When Peter met me at the station in his tattered kilt, and I saw Dick snorting at the train, his coat unkempt and harness mended with string, my heart leapt within me. And when I drove up the farm road, and beheld the snow-covered coils and the stack and Murdo's snowball, and the calves in the sawmill field and Mr Gordon sawing wood in the cart shed, and Flora mashing potatoes in the old familiar mess and muddle of her kitchen, I wondered why on earth I had spent five pounds on my return fare south, unless it was for the exquisite pleasure of seeing our own moist, untidy world again, all the dearer for the short separation.

18

Interlude on Junk and Other Matters

JUNK is the farmer's cross and crown. Every day he is vexed by rampant and unsightly litter, always in evidence, sometimes even at his front door: rotten timber, pieces of corrugated iron, trails of fencing wire, rusty nails, leaky buckets, odd scraps of machinery, and all the miscellaneous lumber that we cannot burn and are too lazy to bury. We let them lie, on the plea that they may come in useful some day; and most of them will. To the shrewd econo-miser, a scrapheap is as good as a cheque, and if he is handy and ingenious, it will give him an excellent chance of showing his skill and saving his pocket. Peter made a large portable henhouse, which would have cost several pounds to buy, out of disused stable partitions, and thatched it with rushes. We converted the open sawmill into a semi-closed hay barn with the help of old nursery thinnings from Rattray's plantation and sheets of corru-gated iron found lying about the farm; not a penny was spent on material, apart from a shillingsworth or so of nails, and many of these were pulled out of boxes or planks. The patent hay tripods are fitted with a set of three corrugated-iron ventilating tunnels, which cost over £1 to buy. We made ours for nothing out of odd sheets. The only disadvantage is that our junk-made appliances were sometimes too queer-looking to be respected as private property. While we were reaping the long cornfield some drain-age work was in progress at Rattray's. From the top of the slope where I was busy binding sheaves, I saw the plumber come up the road on his motorcycle and turn into the steading. He knew where to find me, and I did not trouble to go down. A few minutes later I was astonished to see him emerge with something suspiciously like my ventilating tunnels, which had been lying in the hayfield beside the tripod, ready for the next clearance from the fence. I galloped down the field, yelling at him to stop. I asked who had given him permission to take my corrugated iron. He looked aggrieved, and replied that he never thought anyone would want bits like that, and he needed them to cover a manhole at Rattray's. Stifling a desire to consign Rattray and his manhole to the devil, I politely explained to him that these bits were part of a hay-drying device of infinite value, and he must restore them at once to the place they came from. Whatever cover Rattray got for his manhole cannot have been very secure, for he used to spend Sunday morning before church time in stoning the Belties away from its neighbourhood.

In tidier regions we should not mention fences in the same breath as junk, but at Achnabo most of the so-called fences were mere lines of broken, twisted, and rusty wire strung on bent iron standards and decaying wooden posts, a nuisance and a hindrance, if not an actual danger to stock. At Strathascaig they were a little better, but still in startling contrast with the trim elegance of the house. But there is some excuse even for this. No one who is not a practical farmer of limited means can real-ise the expense and annoyance of making, and even more of keeping up, fences; and by 'keeping up' I do not mean mere maintenance, but the brute physical business of preventing the damned thing from falling flat. For fencing, like a good deal of our work in the West, is one long and generally unsuccessful fight against the law of gravitation. Change and decay in all around we see, and our whole environment – posts, gates, roofs, livestock prices, and spirits – is in a perpetual state of actual or threatened collapse. If only the attraction of the earth went deeper, and the fallen articles, instead of littering the surface, were sucked down into the abyss, it would be much easier to be tidy. As it is, even the most energetic and orderly person can hardly hope to keep pace with gravitation. Whenever I see a row of tottering posts joined by rusty nails or slack wire, I cease wondering why our ancestors clung so tenaciously to common field farming. Who would willingly endure the expense and worry of enclosure, as long as there were obliging old women and confirmed idlers who are ready to herd the animals?

The only sound and permanent fence is a stone wall, or the stone-faced earth-banks so often seen in Cornwall. But not only is the art of the stonemason and Galloway dyke-builder almost extinct, but any kind of work in stone is prohibitively expensive: hence the absurd sight of brick and concrete houses in a country where suitable stone is as common as sand on the seashore. The wire fence, so economical of space and apparently so simple, is, like many another modern gadget, full of snags. It is easy enough to buy so many hundreds of larch posts, so many rolls of wire, and so many bags of staples, and then find someone to hammer in the posts and fix the wire. But in some places the ground will be too hard to give wooden posts a hold, and you must get iron ones, anchored to huge buried stones that take half a dozen men to shift them into place. And everyone who has tried to stretch fencing wire knows that no sooner is one row taut than all the others get slack. And when the damned thing is fixed at last, what then? Immediately the forces of Nature declare war upon it. The action of time and weather rots and loosens the posts. The wire gets slack and rusty, the staples fall out. This gradual decay is hastened by the cattle, who use the fence as a convenience for rubbing, and force their heads through the wires to reach the grass on the farther side. And so in a few years' time the proud erection is tottering to its fall, while the centenarian stone wall, which has never been repaired, will outlive a dozen generations of its modern substitute. No wire fence is effective against livestock unless protected by barbed wire, and this, as everyone declares, is an invention of the devil. But I think that everyone may be mistaken; the devil's devices are more cunning. Like a stupid, officious dog, barbed wire often gets hold of the wrong victim: it tears the farmer's hands and clothes, while the bull uses it as a piquant scratching tool with all the complacency of one who has found a really efficient toothpick.

Gates are nothing but movable fences, so that they share all the pestiferous qualities of the fence, to which are added a whole host of new ones introduced by the idea – I really should say ideal – of mobility. We expect a fence to stand up, but not to open; but a gate should both stand *and* open. It should, if possible, also shut. These hopes are perhaps unattainable in the brief span of a human life, but our incurable optimism prevents us from seeing it, and our special grievance against gates is their failure to live up to our expectations. Long ago in the prosperous 'eighties the farm was provided with a set of iron gates, smoothly swinging on solid wooden or iron posts, and fitted with bolts that really fastened. A few of these are still in good order, but most of them, though the gate itself is intact, have come off their hinges or slipped from the straight, so that they drag in the mud and can only be opened and shut with colossal muscular effort. But iron gates, damnable as they may be, are nothing to wooden ones. A wooden gate will not only refuse to open or shut, but may even not deign to hang together. Small wonder. It is exposed to every kind of force – wind and weather, the pull of earth, butting of animals, crashing of badly steered carts, and tugging of impatient hands. It rarely swings on hinges; loops of wire or even string support its doddering movements. It is repaired, if at all, by nailing new pieces of wood onto the ancient framework, which is much too rotten to hold nails at all, so that the last state is worse than the first. One veteran deserves especial mention. Just as the Jews have a Weeping Wall, so at Strathascaig they had a Cursing Gate. It led from the steading to the hayfield, and through it must pass all the traffic of harvest. It was about nine feet wide, and intended to close the gap between the gable end of the barn and that of the stable. The projecting foundations of these buildings allowed no hold to the posts, which swayed and staggered with every opening and shutting. The gate itself, a lichened oblong frame with upright bars, half of which were loose and too rotten to hold a nail, swung on precarious wire and string, and was fastened by a hazel stick wedged between the upper bar and the wall of the barn. A slight gust of wind would lay it flat in the slime, and so limp and incoherent was the structure that it took two people to set it up again.

Reader, I am afraid you must think us very untidy. And you are right; we are untidy. The climate has a good deal to do with it. Everyone knows that in the clear dry air of North Africa there are Roman remains in a quite remarkable state of preservation, and the traveller in India may see cities which have been deserted for nearly 400 years, and yet look as if they had been built yesterday. But here it is very different. Wind, rain, and the rampant vegetation of a damp soil are incessantly busy in destruction. Even inhabited houses, unless often painted and repaired, soon fall into decay, and everything exposed to the weather rots and rusts away to nothing, so that a short period of neglect will turn the tidiest steading into a rubbish heap. Yet there is no place on earth where implements, tools, and harness are so much left to take their chance, where so many shiftless and ultimately labour-increasing devices are employed to put off the evil hour of systematic repair. String is the Highlander's panacea. I keep some in most of my pockets, as well as a few nails, which can be driven in with any good stone that happens to lie handy. The strict person from the South will say that since everything disintegrates so quickly, there is all the more need of care. But the soft climate suggests another argument. Since everything disintegrates so quickly, why wear yourself out in a vain struggle with the inevitable? Leave the barn unpainted, the plough in the furrow, the scythe in the bracken, the bridle in the stable gutter. In a few years' time these things will all be finished with anyhow. Why bother? At one time I used to wonder why well-to-do immigrants who had taken sheep farms in the West tolerated a state of shiftlessness and neglect which they would never have suffered at home. A little experience taught me that it is futile to swim against the stream; the man who tries Lowland farming in the Highlands will die with an empty purse and a broken heart. String-tied harness and rusty implements will serve the turn of ploughmen who, whether from excess of laziness or of saintly detachment, take no pride in keeping things up to the mark. The alien settler has no need to keep up appearances. The natives are not impressed; indeed, they probably pity him for being the slave of so many stupid little fads. The Highlander has lost his political, but not his personal, independence; and the Sassenach will be wise to abandon himself to the stream of tendency, and carry quite a lot of string in his pocket.

The prize gateway at Achnabo was in a field known as the first loch park, much used for the milk cows in summer. The official gate was of iron, and enormously heavy. The posts had become too rotten to hold the hinges, and unless handled with the greatest care the whole gate would come crashing down on your feet. We wired it up permanently, enlarged an incipient gap in the fence at the corner nearest home, and fitted it with a slip-rail about fifteen feet long, made out of a young tree, running through wire loops. Between this improvised barrier and the road was a wide ditch, and beyond the road the loch, into which the field drains discharged. The level of the loch had been raised to increase the village water supply, so that in wet weather much water flowed backwards up the drains, making the ditch a deep and oozy canal and the edge of the field beyond a sodden morass. The cattle would plunge and flounder across, and the cowherd must wade or jump the ditch to secure the rail behind them. The next field to this was bounded on the west by a strip of woodland, which had suffered severely in the 'memorial gale' of 1928. Most of the fence was wrecked by falling trees, and the ruin completed by men sent to cut wood, who dragged logs out across the wires without troubling to make a proper gap. There were hundreds of acres of woodland about Achnabo, untended and uncleared; there should have been a forester in charge of them, to keep the place in order and prevent the random cutting and hauling of timber. Without supervision, the line of least resistance was always sought. Trees nearest the road were cut, and at a level which enabled the men to saw with straight backs; thus the thickest part of the log was wasted and an awkward stump left behind. Brushwood rotted ungathered and unburnt to cumber the ground and entangle cattle roaming in through broken fences.

The whole estate was rapidly approaching that stage of dissolution when furtive raiding becomes open piracy, and ends at last in systematic squatting and appropriation. I should not blame anyone for taking advantage of these conditions. The landlord was away for half the year, and no properly responsible person was left in charge. Everything wore a sad, neglected air, as if inviting help and attention from outside. I never wandered through the tumbled woods and derelict pastures without longing to clear and drain, to build wooden houses and enclose holdings. This would never pay. Economically, Achnabo and its woods should be annexed to Strathascaig, to support a few more sheep and rough cattle, and perhaps a man to look after them. There is no doubt that the large sheep farmer with plenty of capital can make the most productive use of the land, and from this point of view, the Highland clearances were justified. But as a maker of employment, he is most unsatisfactory. The Laird, with nearly 6,000 acres, employed only one shepherd and a little casual labour. To settle the maximum number of people on the land, an extension of the crofting system would be necessary, but it is very doubtful if such crofts could be made self-supporting without subsidiary work of some kind. Even more questionable was a recent proposal to settle unemployed men on the estate. Highland crofting is a life of its own; to the native or the specially adaptable outsider it has many advantages and attractions, but I cannot think that it would appeal to the average industrial worker, and still less to his wife. There seems no future for the place beyond a gradual disintegration and return to the wild. The thought makes me sad, if only because we ourselves have shared in the work of defending a few hardly won acres from the encroachments of triumphant Nature.

The sight of derelict agricultural land always gives me pain, and not merely because I happen to be a farmer. Nature has a particularly ugly way of dealing with the earlier stages of her conquest.

Stripped of cultivation, the field nourishes an ignoble growth of weeds, fouler and more rampant in proportion to the richness of the soil they have usurped. The useful crop is gone, and the beauty of the wild not yet established. And since agriculture is the natural and basic work of man, to lose our hold on the land is the sign of a fundamental and irremediable weakness. The spectacle of a derelict industry grieves us with the thought of those who have lost their work, but in the ebb and flow of industrial enterprise, men displaced in one trade may often find employment in another; whereas the wholesale failure of agriculture would in the end bring ruin on our country. Britain's former industrial supremacy is passing into the hands of those who can supply cheaper labour and raw materials easily accessible, and there are many who think that before very long we shall be reduced to a third-class agricultural nation. The process of adjustment will involve much hardship, but I for one shall not be sorry. Necessity will force us to make the utmost use of our equable climate and excellent grassland for the intensive production of food, just as it forced a purely agricultural country like Denmark to organise its only productive industry on scientific lines. In Ireland, even the quarrelsome and individualistic peasant has been persuaded to co-operate, because he is shrewd enough to realise that without a prosperous agriculture his country must cease to exist. And what is true for Ireland today will probably be true for the rest of Britain tomorrow.

The decline of an industry and the reversion of a whole district to agriculture can be studied in West Cornwall. A hundred years ago the country was full of tin mines in working, and supported three or four times as many people as it does now. Ruins of mine stacks, engine houses and miners' cottages, abandoned shafts and galleries, are to be found on almost every hill and headland. Then came the discovery of cheap surface tin in Malaya, and the business of blasting one's way through granite and boring under the sea, with the ever-present difficulty of keeping the workings clear of water, became prohibitively expensive; the mines closed one by one, and the surplus population, with characteristic enterprise, sought new work and homes abroad. Those that remained were absorbed in fishing or agriculture, and the Penwith peninsula is now a land of peasant proprietors engaged in dairying, whose yellow Guernsey cattle graze round the overgrown remains of an outworn industry. The lure and dazzle of speculation has gone, and these people work long hours for a very plain livelihood; but there is no poverty, very little sickness, and still less discontent. The conversion of industrial Lancashire to agriculture would involve a far greater upheaval and confusion than this, for the social machinery is more complex, the people affected more numerous, and the opportunities abroad much fewer. But it may come, whether we like it or not. Industrialists are always faced with the possible discovery of a quicker and cheaper process of manufacture, or of some substance which will make their products superfluous. The decay of industry alone will not help the farmer, since it would rob him of his chief market, though perhaps less in these days, when the urban housewife buys Argentine meat, New Zealand butter, Australian eggs, Danish bacon, and American tinned fruit. There is no practical road to happiness. If there were, it would probably involve a drastic reduction of the population, so that we might once more become a self-supporting agricultural community; and this seems no longer possible. But perhaps I may be mistaken; at any rate I sincerely hope so.

With a little more leisure we could have produced our own clothes. Some years ago I had learnt handweaving from an English friend, and hand-spinning from a servant who was a native of Harris. My own wheel and loom, which had been used at Strathascaig for turning the Laird's fleeces into tweed, were stored in the lumber room at Achnabo: but all these things take time. I have heard people complaining of the price asked for genuine hand-woven tweeds – usually 7s a yard, but often less; they would be sorry to produce tweed themselves at this figure. Having myself done the whole process, carding excepted, from the sheep's back to my own, I can claim to know something about it.

In this district sheep are not washed before clipping, and the fleeces are packed as they are, full of dirt, grease, tangles, and scraps of twig and heather. They must be washed very thoroughly with soft soap and ammonia, often in six or seven waters, and then hung up to dry. The handling and wringing of fleeces is heavy work, and drying them a slow business, for they must be kept away from livestock and from places where fodder is stored. Scraps of wool have killed many calves; they lodge in the stomach, where the hooked fibres gather food around them, until little by little a hard ball is formed which causes congestion and finally death. When the fleeces are quite dry, the worst tangles must be teased out by hand before carding. Most spinners now send their fleeces to be carded at a mill, but a few still do it at home, and the results are more pleasing and durable, especially where very soft wool such as Shetland is used. Dyeing is usually done after washing and before carding. Herself was an expert dyer, and we made many experiments with various vegetable dyes; but the commonest and perhaps most effective is that obtained from crotal – the greyish-green lichen that grows on rocks and old stone walls. A fire is built outside, and a large cauldron or boiler set over it. This is filled up with wool and crotal in alternate layers, pound for pound, and allowed to boil for about two hours, when the wool is taken out and the scraps of crotal shaken from it. The colour obtained is a rich reddish brown, which when mixed either in carding or in weaving with other colours, is the foundation of most of the well-known Harris tweed. The wool is dried once more, and then carded. The effect of this process is to tease out all irregularities and shape the mass of wool into long thin rolls that can be easily handled by the spinner.

To the beginner, spinning is the devil's own job. It takes about a fortnight to learn to spin with competence. Like many other things it cannot be taught; you must just practise till you get the knack, and try to master your temper. The wheel must be controlled with your feet and the yarn with your hands; the problem is to work them together, and the worst moment is the joining of one carded roll to another without stopping the wheel or allowing the yarn end to run out of your hand and on to the bobbin – a thing it will do every other minute; and this means tedious re-threading with a bent hairpin, and fury and despair. Or the wheel will suddenly run in reverse, with tangle and confusion, or it will slip its band, or refuse to turn the spindle at all. If the wool goes onto the bobbin too quickly, it will be thick and lumpy; if too slowly, it will overtwist and break. When the bobbin is full, the wheel must be disconnected and the yarn wound off into a ball for convenience in weaving. If knitting wool is required, two balls are taken, each put into a separate pail, both ends led to the spindle, and the wheel worked in reverse: the separate strands will thus be twisted into two-ply yarn.

Weaving is a far more skilled and complicated craft than spinning, and very few hand-looms are now to be seen on the mainland. Most local spinners send their finished yarn to a handweaver or a mill to be made into tweed. But if the cloth is to be woven at home, the process is roughly this. The length of tweed desired – say twelve yards – and the number of lengthwise or warp threads – say eighteen to the inch – is decided upon. Sufficient yarn is taken to the warping-board, where it is wound on spaced pegs in such a way that when removed it will lie in the form of a woollen serpent twelve yards long, consisting of 472 separate threads arranged in a *definite order* – that is 18 × 27, for most hand-woven tweeds are twenty-seven inches wide. The threads are then taken to the loom and passed *in the same order* through the gaps in the reed, a comb-like contrivance fixed on the beater, which keeps the warp threads evenly spaced, thus controlling the width of the cloth. If this order is once lost, the result is complete and hopeless chaos. After passing through the reed, the yarn is threaded, *still in the same order,* through the heddles at the back of the reed. These are loops of wire or twine strung on pairs of sticks, which are connected with the treadles, and work in twos, fours, or sixes according to the weaving – plain or patterned – that is desired. The ends are then tied to the roller at the back of the loom, and the weaver, helped by an assistant, slowly rolls the whole length of warp threads on to the back roller, leaving just enough length of yarn in front of the reed to tie the ends to the front roller. During the process of rolling, all threads must be kept at an even tension or they will sag or break. The warp is now ready for weaving. The rest of the yarn, or weft, is wound on a spool-winder into small bobbins for insertion in the shuttle. The actual weaving of the material is done on the space between the front roller and the reed. At each movement of the treadle the weaver raises or depresses a set of warp threads, and through the tunnel or 'shed' thus made, the weft thread is thrown in the shuttle, and beaten hard down on its predecessor, thus making the cloth. As the weaving proceeds, the rolling process is reversed, until, when the piece is completed, the whole length of warp threads will have been gradually transferred from the back roller to the front. The tweed is then taken from the loom, and the loose ends roughly bound over to prevent fraying. It is then washed to remove the grease left by the carders for ease in spinning. When nearly but not quite dry, we smooth it flat on a big table and roll it tightly and evenly on a wooden roller, upon which it is left to dry out completely. From this bald and imperfect description the reader will realise that the home producer of tweed deserves 7s a yard. The price of wool has little effect on the price of tweed; about three-quarters of a pound of Cheviot wool, now 1s 1d a pound, goes to a yard of fairly heavyweight tweed, and the cost of the raw material is, as in so many other cases, very small in comparison with the cost of labour.

These tweeds give very long wear. They are not as hard as certain of the closely beaten mill-woven cloths, but they have a way of moulding themselves to the figure, and of getting shabby in a graceful and gentlemanly fashion. Vegetable dyes fade into pleasant hues; dirt blends well with crotal, and frayed edges attract no more attention than the fringe of a kilt. You may spend a good deal on your suits to start with, for they need good tailoring, but you can wear them till they fall to pieces without losing caste. There is a dark shed in the steading at Strathascaig where the

Laird has a mausoleum of old tweed coats, which are kept for sheep dipping or for presentation to tramps. But they are so well worn that the tramps generally decline them! The following incident will show at once our prevailing economy and the durability of hand-spun yarn. I bought some hanks of fawn-coloured knitting wool at the Highland Exhibition, and gave them to an old woman to knit into stockings for me. She made them far too large, and they were returned for alteration. After that I wore them incessantly until the feet were done. I then had them re-footed once and twice. The wool of the legs was still as good as ever, so I unravelled it, wound it into balls, and put it away. A year or two later I had an order to weave some striped woollen cloth for cushion covers. My colour scheme required the introduction of a narrow fawn stripe, but I had no fawn hand-spun wool. Then I remembered those old stockings. The yarn was hastily wound into bobbins and took its place in the design, which was much improved by its presence. Some time later, when my customer was admiring the hard-wearing quality of the cushion-covers and the beauty of their colouring, I told her about my old stockings, and she, being a Scot, laughed heartily!

19

November: Port after Stormy Seas

I BELIEVE that most farmers, in their hearts if not openly, prefer winter. If asked for a reason, they will probably say, because there is less work. This is a sensible answer and not as lazy as it sounds. We may regard work as a curse, a duty, a privilege, or a pleasure. It is all four; but however great the delight we feel in working, nobody wants to toil without a break. The best moment is the last, when straightening our backs we contemplate our handiwork and say, 'Well, that's that, and now, thank God, to rest.' This, according to the author of Genesis, was the attitude of the creator on the first Sabbath, and is surely good enough for us. On Sunday morning, the traditional farmer takes a walk round the farm, watching his crops and beasts, and thinking about them. There is something quiet and spacious about this employment of leisure, and it may be profitable also; for these are no idle speculations – they lead to some end. A chimney-sweep once told us that he made more by scheming on Sundays than by working on week-days; and I believe it.

However strenuously we may work for a season, rest comes at last; and the beauty of the West Highland climate is that we have not only sabbatical days but a whole sabbatical month. In the south November has acquired an evil reputation. It is a Cinderella of the seasons, hateful with dismal fogs and gathering darkness. In Brittany they call it the Black Month, perhaps on account of its short days and dim light, but more probably because 2 November is sacred to the souls of the departed. But since pious Bretons believe that the souls of the departed are at rest in God, November should not be called a black month of mourning but a grey month of quiet recollection, soft as the lingering twilights of the north. In these latitudes no one ploughs or sows in late autumn; they sit and think. With any reasonable luck, you have secured your hay, corn, and potatoes. Your cows, calving in spring and suckling their young, are by now dry or nearly so, and as the grass remains green and palatable far into the winter, they do not yet require food indoors. Most of the calves are either sold or weaned. People have ceased to bring cows for the bull. Shepherds have finished their last dipping, and are making leisurely preparations for sending out the tups to the hill. Weeds, as far as this is ever possible in so moist and mild a climate, have ceased growing in the garden. It is getting dark in the morning and no one rises very early. You saunter over the russet hill, visiting out-wintering cattle; you potter about the steading, cut and cart a little wood for the fire, and curse at the hens because they also feel sabbatical and have ceased laying. For the rest, you just sit and think.

This year above all, when a calm and beautiful November had followed three months of unbroken storm, we felt more than usually eager to enjoy the truce of God. We blinked in the mellow unaccustomed sun, like the battered crew of a ship that has made a winter passage round the Horn, and now lies in some quiet harbour to refit. It seemed strange that there was no longer any need to scan the sky, to bolt our food, and rush forth to secure coils or hand-ricks. We ceased to tap the barometer or study meteorological charts. The tension was relaxed and we became thoroughly lazy. Forks were lying in the stackyard, the traces were still on the sledge, and the ropes remained on the fences till New Year's Day, when, fearing a visit from the Laird, we went round the fields and took them in. For the sake of economy, and partly also for peace, I dismissed Flora with a reference, in which I praised her honesty and loyal service, and began to turn my attention to cookery and housework. Peter now did nearly all the outside work, and it seemed quite pleasant to keep warm and dry indoors. Life became quiet, peaceful, and orderly. We had better and more varied food, and long evenings by the fire for reading and study.

On one of these halcyon days we strolled up the road to cut bracken. It was pleasant to think that we need not hurry or worry; we were not working for food, but only for bedding, and if the frost came and crumbled the bracken to powder, our Spartan cattle could do with rushes or go without. I had a fork, and Peter carried the scythe, with which in the past few months he had cut so many swathes of soaking grass, and with it the well-worn stone, now transferred to the kitchen for sharpening knives. We climbed the little brae above the farm, past the old bath that served as drinking-trough, past the little wood where the Belties lay on stormy nights, and pursued our way by the stubble of the cornfield and the trim white and green of Rattray's house, to where the road emerged from tall trees upon the quiet levels of the loch. The place we had chosen for cutting was upon the Laird's land, a little way above the road, but conven-ient for hauling bracken down to a cart waiting below. The day was mild, sunny, and utterly still, as if that pitiless wind had gone to vex some other planet, leaving profound and everlasting calm. The nearer slopes were tawny with birch and bracken; on the eastern hills behind the shepherd's was a rich purple bloom like plums. Below, the loch lay still, with faint flutings as of antique glass, reflecting every detail of shore and sky, until a troop of mallard drakes rose with a whirr, shivering the surface of the water and the spell of its silence. We set to work, while the two dogs, plain enough in detail but superbly graceful in movement, chased each other with worrying noises in the rustling dead fern. The bracken, flattened by weeks of hammer-ing rain, was badly laid and hard to cut. As Peter mowed, I gathered it into heaps with my fork, and before leaving the place we made a small stack of the harsh crumbling stuff and bound it down with rope to await the coming of the cart.

The sun had set, but a full moon was rising behind the rocky heights at the head of the loch. On its northern and western shores, the water brimmed to the foot of enclosing hills, whose gentle slopes were clothed with bushy heather and young birches; but at the eastern end, from the mouth of a steep and gloomy glen, a swift burn came forth, and after flowing for some distance through grassy flats, discharged its hurrying waters across a shallow bar of gravel, fringed on either side by tall reeds; and then, traversing the placid level of the loch, it plunged down a steep wooded gorge, skirted the village crofts below the bridge, and found the sea in one of those many lovely bays that face the sunset and the hills of Skye. At the upper end of the flat, under the shadowing rocks that hid the sun in winter, stood the shepherd's house, with its byre, haystacks, and croft. It was nearly a mile away, but the last light of day was shed upon its whiteness, and the rising moon stooped over it from the east. I could see Duncan with his dogs driving home the cows to be milked and his wife coming out with her pail. It was a lonely place; the childless couple were getting old, and there were no neighbours near, no change, no amusements; only the daily round of work, food, and sleep, the murmur of running water, the soughing of the wind against the gable, the cry of sheep upon the hills. As I watched the white cottage and the tiny figures moving, a pang of envy seized me. To live in peace year in, year out, in some remote and beautiful place, with work to hand, a loving mate to share it with – what more could man desire? – or woman either?

The moon rose full and clear at our backs, and dimmed the fading glory of day. We turned the corner by the Belties' little wood, and from the upper gate looked down upon the farm. A thin smoke curled up from our chimney; the fire was still in, waiting to be stoked for tea. An answering smoke, thicker and better tended, rose from the Gordons' cottage. Mr Gordon was pottering round the henhouses, shutting them up for the night. The cows had been grazing all day in the field behind the stable, and were grouped at the gate, replete and happy, sleepily chewing their cud. The stubble and aftermath were clean and empty; only a few heaps of shrunken mouldy hay, torn from the fence and left to lie, reminded us of the warfare accomplished. The fields wore a quiet collected look, as if meditating their past fertility and future hopes. The southern end of the sawmill was packed with corn, so full that straw was pushing out at the open side, a delight to the eye and temptation to questing beasts. Dear Achnabo! What person, I wonder, will stand here next year and look upon its deep autumnal peace, and bless it? Probably no one. 'Life', an old Cornish farmer once said to me, as we leant over the door of a pen and looked at his calves, 'Life be nothing but meetens an' partens.' It is just that.

At the steading we divided; I returned to the house, and Peter went down to fetch the cows from the field. Mr Gordon had finished with the poultry, and was sitting indoors with the newly arrived newspaper. Since Flora's departure I had become more feminine in my occupations, and did not even go out to milk; it was now Peter's business to take the cattle home from pasture, shut up the bull, milk four cows, and feed the calves. The providing of tea had become more complicated, as the older hens had ceased laying and the pullets not yet fairly started. Also the supply of preserved eggs, recklessly used by Flora to stay the appetites of haymakers, had lately come to an end, and under these conditions the cook's life is not worth living. What is bacon without eggs, or sausages either? Macaroni, perhaps – macaroni cheese. Luckily Peter could eat anything, a charming trait in man or beast. Macaroni it should be. Then I remembered that there were kippers in the dairy. Peter had gone to the town to get a jacket fitted. The tailor had not begun to think of making it, so he strolled down to the pier, saw a boat coming in with herring, let down a wooden box and hauled up over a hundred for a shilling. These he took with him in the train, and wheeled up from the station in a barrow lent by Flora's uncle. Some were given away, some eaten fresh, some salted, others kippered, and the rest left inadvertently outside the house, until they began to annoy our noses and upset the dogs, when they were formally buried.

I suppose that conscientious people do not keep kippers in the dairy. But our dairy was very large, and the dairying operations confined to the dumping of pails and straining of a few pints of milk for the house. So it had become a kind of storeroom, larder, and tool-shed, and contained fish, bacon, fowls waiting to be plucked, a bag of calf meal, another of shells for the hens, a few eggs, a wringer, a grindstone, a bag of nails, and on the days when the baker called, a few loaves and a dozen of buns. It was dark when I went in, and turning to the shelf where I knew the kippers were spread, I was amazed to see a row of glowing points. Each kipper had a few phosphorescent spots on the inside. They were perfectly fresh.

I lighted the lamp, stirred up the fire, and put the pan on the stove. While the fat was heating I began to consider what we would do tomorrow. The supply of wood was getting low; Peter had recently set and sharpened the saw, and I planned to take dinner and spend a whole day in the woods, so that we could cut three or four cartloads at once. The idea pleased me. Wood-cutting is one of the best jobs of the year; the work itself is pleasant, and the thought of good fuel for nothing delights the thrifty soul. The woods were all around us – our own pines, larch, and spruce, the oaks beside the road, and the Laird's birches beside the loch. The fat began to smoke; I put the herrings in the pan, and watching them mechanically, fell into meditation.

Down in the murky depths of consciousness, where civilised people do not care to probe, lest they should find something discreditable to their pretensions, lie many primitive terrors, dormant perhaps, like sleeping crocodiles, but horribly vital: and among them is the fear and hatred of trees. Not of isolated trees, nor of groups of them, but of trees in mass – serried armies, stifling crowds, trees that devour the daylight, hide the stars, drain the air of freshness, and harbour unseen enemies. On the open hill there is space and freedom, light to see by, air to breathe, room for flight or conflict. In the woods all is secret, suffocating, dangerous. As the townsman feels when the Tube train stops between stations, or the ship's passenger when he is battened down below in bad weather, so do we feel in an unknown wood at dusk. We are blind, choked, without sense of direction. I have never seen a tropical forest or a mangrove swamp; whoever has, must have supped full of horrors. But only to read about them, even in the uninspired pages of the average book of travel, is quite enough to make my hair stand on end, as it would never do if I were reading about a mere ghost or werewolf. There is more terror in Nature itself than in all tales of the supernatural that were ever told. The plain facts about insects, even in our temperate regions, can give us keener thrills than were ever provided by man's most fertile imagination.

In Britain the forests, like the insects, are of a milder order. The nearest approach to horror in trees is to be found in the ancient dwarf-oak coppices of the south-west, like the famous Wistman's Wood on Dartmoor. These stunted trees, whose gnarled limbs, twisted by age and storm, are covered with tufted lichen, resemble nothing so much as a crowd of hairy greybeards – old, incredibly old – who claw at the passer-by, trying to make him stop and listen to their story. Of course this is all imagination. Lop off a few rotten sticks and light a fire: that will dispel your morbid fancies. But the sticks are damp and mossy; the fire sizzles, sputters, and goes out. Darkness begins to fall. Is it all imagination? Once as a child I was taken to a village in the Chiltern Hills, where our chief occupation was walking in the beechwoods. I would have given much to avoid these walks. Not that the woods were grisly or dark – far from it; the straight smooth trunks had a rare beauty, while the absence of undergrowth and comparatively wide spacing of the trees gave the illusion of a huge hall with innumerable pillars. But all the same there was some horror present. Whether it was the naked glimmer of the light outer stems against the interior gloom, or the sinister rustle of dead leaves underfoot, or the mere bareness of the ground beneath the trees, as if they possessed some poisonous influence, deadly to alien growth, I could not now say; but from that day to this I have never cared to walk in beechwoods.

In the Australian bush, trees are terrifying from their mere number, persistence, and monotony. The eternal grey-green gum trees, which never shed their leaves nor change their colour; mile upon mile, league upon league, inspire uneasiness, fear, and in the end loathing. Crowding in millions, they press upon every clearing, and the settler's truceless warfare is not only with drought but with forest. Nor does killing remove the enemy from sight, for the usual practice is to leave the ring-barked trees to die where they stand, so that the paddocks are full of bleached and hideous skeletons, useless for shade or shelter, on whose topmost boughs are perched a crowd of harsh and unmelodious birds. To be lost in the bush, where each new tract exactly resembles the one before it, where you cannot see more than 200 yards ahead, where there is nothing to steer by but the sun fitfully spied among the tree-tops, would bring a deeper sense of terror and desolation than to be set adrift without rudder or compass on a starless and fog-bound sea. I had always supposed that, apart from mountain and glacier, the grim island of Tierra del Fuego was a waste of bog and barren moor. But nothing so open and kindly. Leagues of impenetrable forest, oozing with water and foul with rotting timber, fringe these unfriendly shores, extending from the snow-line to the very edge of the tide. A fearful place.

If I have dwelt so long upon the awe-inspiring qualities of woods, it is only to emphasise the pleasant and jovial labour of the woodcutter. For wherever the merry sound of axe and saw breaks in upon its solemn silence, the gloomiest forest is robbed of half its terrors. Trees are no longer hostile armies, but stores of timber to warm and house the worker. Here, within a short distance of the farm, are woods of various kinds, old and young, natural and planted. The ancient Scots fir, remnant of the primeval forest, whose savage grace tempers the harshness of many an inland glen, does not seem to grow so near the sea: the natural woods were mostly birch. Many of the hills of Wester Ross have a cliff-like appearance, the upper part being in the form of a sheer or even overhanging crag, while the lower slopes, scarred with water-courses and screes of loose stones, fall steeply to a loch or boggy flat at the bottom. These slopes are often clothed with ancient twisted birches, grotesquely sprawling in every direction, hairy with lichen and adorned with jutting clumps of orange fungus. Beneath the trees the ground is incredibly rough, a jumble of fallen stones and boulders, some bare, some furred with crotal, some loose, some fixed, some round, others jagged, but all unpleasant to walk among. Between the stones are holes and cracks of unknown depth, veiled with treacherous moss or heather, and full of water, slime, or rotten twiggery that crumbles at the touch. The whole place reeks with moisture; a hundred small springs ooze from the crevices of the stones, and seep down-hill to feed the bogs below, their courses marked by fan-shaped patches of sphagnum moss and tussock grass. Except that there are no thorns and no poisonous creatures, I can hardly imagine a worse place to walk in, especially if you are in a hurry or it is getting dark. I often wonder what will be the end of this kind of wood. There appear to be no young trees; it is a senate of hoary old men, and when the last of them has crumbled into dust, the whole brae will be a waste of bog and stones. On the gentler slopes, where the hills first rise from the glen bottom, and espe-cially in the gullies which the burns carve out for themselves, are woods and coppices of young birches, whose straight smooth stems, silvery or reddish brown, seem to have little in common with the gnarled hairy limbs of the elders. In winter the whole mass of branches and twigs glows with a lustrous crimson-purple, impossible to paint or describe. Perhaps it might best be imitated in a shot web of cloth, where crimson and purple yarns were thrown across a warp of natural black. For the undyed wool of a black sheep has that strange luminous darkness, a darkness that is not black, which is the foundation of the birches' winter colour-ing. And in late November, when a few orange tawny leaves, touched by the level rays of the sun, are still hanging on those dark yet glowing boughs, you see a combination of colour beyond the dreams of any artist or weaver. There is something friendly about the young birches. They grow closely enough to give

272

warmth and shelter, but not so as to exclude the sun or choke vegetation. The shade they cast is dappled with sunlight, and the ground beneath them is mostly covered with soft green grass, which is rarely bleached by frost. In every gully a small burn tinkles over the stones, fringed with ferns and smooth-stemmed rowans and silvery polished rods of hazel. The place is full of life, elusive, half-seen: the whirr of a rising woodcock, shy roe-deer peeping among the trees, the white flicker of a rabbit's tail. Tufts of wool caught in the lower branches show the value of these coppices to the shepherd and his flock; but any excess of tangled young growth must be cut, gathered, and burnt, or the sheltering wood may become the grave of ewes heavy with lamb.

My first experience of wood-cutting was at Strathascaig. In the fall of the year we would set forth – no timber gangs with vast tackle and huge horses, but amateurs with axe and saw, bent on cutting blocks to cheer our hearths; for though there was enough even of fallen trees to warm a city, the roughness of the ground, expense of labour, and difficulty of transport would have made it an unprofitable venture. There is nothing pleasanter than a fire of birch blocks, and a day in the woods makes a welcome change of labour. We would wait eagerly for the leaves to fall and the sap to go down. With a tinker wife's passion for sticks, I began to wonder if the season could be anticipated by cutting a few dead trees. One fine afternoon in early November I went out to prospect.

Just at the bend of the road, where it left the glen to run westward above the shores of the loch, was a planted wood of larch and Scots fir covering the steep slope between road and railway. The trees were very close together; many were dead or dying where they stood; others, mown down by the gales, were lying prone or half-supported by the branches of growing neighbours. These last would be fair game for our axes; so would the dead or dying firs, most of which were now too soft to be used for fencing. They had the great advantage of being far lighter to handle than the green birch; but on the other hand, the birch grew above the road and could be rolled down, while the fir had to be hauled up. The ground was horribly rough; underfoot lay a mass of twiggery of all sizes, tempting enough, if only it were dry, to the gipsy or camper. Fired by some primitive feminine instinct, I gathered an armful and placed it on the road to wait for the next passing cart. Then, having made a mental note of various likely trees, for I could do no more till Willie joined me with the double saw, I sat down awhile, until the wet soaking through my kilt set me once more on the move. The kilt is useful for sitting in bogs, just as it is when attacked by strange and savage dogs, since it interposes so many thicknesses between your skin and the enemy.

I saw the stems of fir and larch, straight as pillars, some dark in shadow, some fired to red by slanting sunbeams, and through the stems a glimpse of loch, blue as the fabled Mediterranean seen through olive groves; and beyond that, the bronze and purple of distant hills, and above the hills, clear sky and slow billowing clouds. The very stillness of the air gave much to hear, since nearer sounds were more subdued, and distant ones had wider range. The habitual soughing of pines and lapping of water on the railway embankment were hardly heard; the alarm note of birds, so frequent when I first came into the wood, had ceased, and a few dry days had silenced the smaller burns. I could hear the click-clock of a fisherman's oars as he rowed out to the haddock bank, and the 'chut! chut!' of a blowing porpoise. In the township across the water, women were talking and calling their hens. Farther away, someone was pulling down a boat with a harsh grinding of pebbles. Half-way across the loch, a mottled flotilla of eiderducks were mingling their quacking noises with a liquid crooning note, which sounded like a polite but rather scandalised person saying softly, 'Oh no! oh no!' On the hill, a shepherd called his dogs. Far below me, but not far in distance, a surfaceman passed along the line, tapping the wedges into the 'chairs' with his hammer. This near, practical sound roused me from dreams. I turned homewards. At the corner I met the Prophet, who had been helping the shepherd to gather sheep for the dipping. He pointed to one of the highest and narrowest ledges of the crag where a ewe was stranded in a place impossible of access. I had seen the ewe before, but we spoke of it for some time; he finally remarked that he must finish thatching a cornstack in a township more than three miles away. It was nearly four o'clock. I reflected that he would not have much time. But never mind: I entrusted him with a message to Willie to come up next day if the weather was good.

The weather was very good. Soon after nine Willie and I set forth with axe and saw. On closer inspection the fallen trees turned out to be larch, too sparky for fuel. This was a heavy blow, for they were both nearer the road and easier to cut. We set to work on the standing firs. Except for the difficulty of finding a firm foot-hold, and the danger of getting their branches entangled with other trees as they fell, the sawing and trimming was easy work. But hauling the sawn lengths up to the road were quite another matter. The logs were light for their size, but the slope was not only steep but full of loose stones, holes, strips of bog and rotten stumps, the whole covered by a treacherous layer of decaying twigs. At the top was a slack wire fence, over which the wood must be lifted, and half-way up the crumbling embankment of an old drove road, making an awkward step for burdened climbers. By a combination of carrying, pushing, and dragging, the logs were hauled up to the road. One difficulty was that the best way up was not always plain from the bottom, and that it was often impossible to change one's course in the middle. From the place where the last tree was cut, the road was completely hidden by a mass of thick laurels. Gallantly we ascended, carrying between us the largest and heaviest of our logs. At last the laurels were reached; but beyond them was no wire fence, but a high stone embankment, which rose like a cliff wall between us and our goal. 'I often wonder', said Willie, resting his end of the log on the ground, 'why Napoleon dragged all those cannons across the Alps.' 'It would have been much better for the world', I replied, laying down my end of the log, 'if Napoleon had stopped quietly at home.' This was the end of our wood–cutting below the road.

I can remember many days in the wood with various companions: with the shepherd, who despised all labour unconnected with sheep; with a freckled girl of sixteen, whose father had trained her to help him in all his forest labours; with a long-tongued, short-sighted boy who did most kinds of work with equal lack of skill, and assisted his crooked vision with a pair of spectacles picked up at a sale. But the best companion of all was Willie, with his dash and energy, his resourcefulness, good manners, and fund of amusing talk and woodland lore. So it is to memories of wood-cutting with the Scout that I turn most readily.

I remember one day in mid-November, when we went to cut our first load of birch. The previous night had been clear and cold, and in the open there was still frost on the grass and a powdering of snow on the higher hills. But among the trees were pockets of warm air, as if stored from one mild spell to the next. For not until the New Year is there any prospect of continuously cold weather. There was every sign of a coming change. The faint easterly breeze had died out, and though the surface air remained calm, there was a strong 'carry' from the south. A filmy veil, etched with faint undulating lines, radiated from the coming quarter of the wind. The last week or two had wrought a great change in the woods. Strong gales had hurried on the falling of the leaves; the ground was soft with a carpet of orange larch needles, and the coniferous trees seemed to dominate the forest. But our business was not with them. We walked on under drooping unpruned branches, through mud and puddles large enough to have a better name. On the right hand was a low stone wall with a wire fence on top, breached in places by big boulders that had crashed down from the rocks above; on the left, a steep slope covered with birches of various ages, interspersed with rowans, hazels, and an occasional holly. The wood became more open and the slope less steep; frequent stumps, some weathered, others freshly cut, showed that for many years this had been a favourite resort for wood-cutters. It was the place for us: we began to look for suitable victims. The trees selected must be large enough to provide substantial blocks, and small enough to be handled by an oldish man and a woman without tackle. They must be placed so as to fall without fouling and be easily rolled or thrown down to the road, free from awkward twists and the multitude of useless little twigs with which the trunks of certain birches are beset. Surprisingly few trees answer to all these requirements, and they naturally become fewer every season. Our usual practice was to cut small trees, which would supply one, two, or at most three lengths for the cart. But this time we had the notion of trying something larger, which, if sufficiently handy to the road, might save us time and labour.

At last we found a suitable tree. Though large, and inclined to be twisted, it stood well apart from its fellows, and was leaning in a direction that would make it fall clear and towards the road. It must have been of considerable age, for its limbs were knotted and lichened, and the smaller branches at the top infested by a twiggy parasitic growth which at a distance resembled the nests of hoodie crows. Under the strain of prevailing winds, the trunk was inclined at a steep angle, making it possible to stand on it and remove some of the branches before severing the main stem. Perched aloft, Willie worked his end of the saw from the tree, while I worked mine standing on tiptoe, balanced precariously on the treacherous ground. Thus we removed five or six branches, and then decided to cut the main stem. It was a tough job. The trunk was thick and the saw blunt. The ground beneath the tree was the usual combination of water, loose stones, and rotten debris: no sooner was an apparently good foothold obtained than it gave way and another must be sought. Standing, kneeling, and squatting were all tried in turn. But we worked on doggedly until the gash widened and a rending crack proclaimed the end. The whole thing staggered and fell with a crash, luckily in the right direction. There is a stimulating excitement about sawing trees which amply compensates the woodcutter for aching back or sore arms. The sharp satisfying smell of newly cut wood, the flying white sawdust, the shrill scream of the saw, the gaping cut, slowly but surely yawning to destruction, the tearing noise when the overweighted uncut fragment begins to give, the shiver of the whole tree before it falls, making you wonder if it will after all turn inwards and jam the saw, and the final triumphant crash at the end, when the white stump with its irregular concentric rings is left bare to the sky.

With his big axe, Willie began to sned the felled branches, while I stood by to roll the smaller ones down to the road. In the intervals of idleness I looked about me. Though comparatively open, the place was very wild. Not far above the woody slope ended abruptly in a sheer wall of rock, from which the blows of the axe were thrown back in double echo. Wisps of cloud, driven by a wind which must by now have been very strong on the tops, poured like water over the edge of the precipice. The frost was all gone; under the lee of this tremendous cliff the air was warm and oppressive. Now and again a stray gust, tepid and smelling of rain, sucked down a gully and set the trees soughing. The ground was strewn with rocks and boulders fallen from above, and one had the feeling that at any moment some toppling projection of the crag might come down like an avalanche, and then farewell to long evenings at the birchwood fire. The place was a haunt of buzzards, and Willie told me that after dark wildcats had been seen crossing the road.

I began to roll down logs. Sometimes they would fall foul of a bush, or jam in a soft pocket; but if one were lucky they would go bumping down the slope till they fell with a final crash upon the road. Willie preferred to balance them on their small end and send them head-over-heels, somersaulting down the brae in the grand professional manner. This sport was not so dangerous as the person accustomed to crowded thoroughfares might suppose. The road was in private ownership, and in a condition to terrify the most reckless motorist. The very occasional local pedestrian or cyclist would hear the noise and be on the look-out; the only people with cars likely to use the road in winter were the doctor and the nurse, and they would have the sense to know that in a place like this anything might happen. If they found lengths of timber lying across the track they could just move them. No one was likely to be so ill that a five-minutes' delay would kill or cure.

The midday goods train had just roared through the nearest cutting, and having the road well blocked with logs, I proposed to Willie that we should haul them to the side and then go home to dinner. We were just slithering down the slope to do it when we heard the sound of an approaching car. It was the dapper new baker's van that smiled once a week upon our solitude, making it unnecessary for us to break our backs and scorch our faces bending over scones on the girdle. The young driver jumped down and cheerfully helped us to clear the obstruction. It was only when the van was once more on its way that we stared at one another with blank surmise, wondering why in the world we had not asked the fellow for a lift. But the long Highland tongue was our friend. Just round the next corner the baker had stopped for a little business and gossip with the post's wife, who had climbed up from the shore to fill an old mail-bag with a week's supply of loaves. Willie heaved his ancient cycle over the wall, the axe and saw followed, and the two woodcutters packed themselves into the narrow front seat of the van. The baker, slamming the back door upon a brightly coloured collection of cakes, buns, and loaves arranged upon sliding trays, mounted beside us and off we went, to be dropped at our gate a few minutes later with a malt loaf purchased by way of thanks.

The same afternoon we went back to finish the morning's work; the next day was devoted to carting the logs to the steading, where the Laird might be induced to cut them up with that most infernal of machines, the electric saw. I shall not say much about it here. I watch it at work with a fascinated horror. It is one of the most useful of possessions, and at the same time a diabolical nightmare. Its strident scream would set the tusks of a rhinoceros on edge. Its sawdust, instead of falling gently like pollen from a flower, is whirled through the air like grit in a sand-storm. Let a loose block touch the blade in motion, and it will whizz across the shed, probably stunning the operator. Let a half-cut block get jammed, and it will dance on the bench like a thing possessed, and then flying in parabolic curve, brain the horrified bystander. Sometimes, when the saw is not working true, it will playfully throw the belt, leaving the mindless dynamo running with nothing to drive. My final advice to anyone is, leave electric saws well alone.

But to return to our carting. Willie went off to look for the horse, leaving me to remove the sides of the cart and fix in their stead a wooden spar at each corner and a U-shaped iron support in front. We hunted the whole place for a rope, and at last discovered the ex-painter of the boat. This would serve our need; and having patched up the harness with a few extra strings, Willie took the cart over the ford, while I cut across the flat on foot to join him later on the road. The loading was a work of art. The logs were of various shapes, weight, and lengths, and must be arranged so as to keep the main weight over the axles, and to the front. They must not work loose or touch the horse, nor must they project sideways, to catch in trees or gate-posts. We hoisted them into position, putting the butt ends to the front: then the load was tightly roped, and off we went, Willie at the horse's head, idly flicking him with a switch of supple larch. The road was narrow, and the loaded cart filled all the space; there was too much noise to talk, so I walked alone in the rear. For all I could see or hear of my companion he might not exist. There is nothing more conducive to meditation of an aimless kind than walking at the cart-tail. I see nothing in front because the cart looms up like a cliff, and blocks the view. I hear nothing anywhere because of the rumble of the wheels. Far from guiding my destiny, I cannot even see where it is leading me. If an abyss were to open under the horse's feet, and the whole thing be swallowed up, I should plunge after it before there was time to think. Without mind or will I follow the creaking thing in front, becoming each moment more like a tinker's dog tied to the axle-tree. I walk on the central dry ridge between chains of puddles linked by canal-like wheel-tracks. My eyes are on the ground: I see the horse lift his feet, revealing for a moment the underside of the hoof with its gleaming shoe, then set them firmly down. Their rhythmical movement fascinates me. So do the firm strong wheels, iron-bound and slanting a little inwards. They press deeply into the sandy deposit washed down by rainstorms, and fling the muddy water sideways out of the puddles. He was a genius who first invented wheels; I wonder when or how he did it. What was the origin of corn? Who first made fire or tamed a horse? These are questions that none of our historians can answer. They can only tell us who founded Rome or who first made laws for the Athenians. More shame to them: for does anyone really care about these things?

20

December–January: Hibernation

WEEK followed week in calm monotonous round, with Sunday supper at Strathascaig as our sole change and diversion. The reader will perhaps be wondering why I have made my sequence of the months begin with February: was the first month carelessly forgotten or purposely left out? Not quite either; but in our life at Achnabo, January is so nearly a mere prolongation of December that I have defied the calendar and telescoped the two into one. An old friend, who often used to spend his August and September in the Western Highlands, once exclaimed, 'Live there in winter! Why, all life is suspended!' Suspended is the right word, but the sense of the remark is all wrong. According to him, the period between one summer and the next is a sleep and a forgetting, a dreamless blank, and at the springing of the first flowers we take up the thread of existence exactly where we left it when we saw them die. All life is indeed suspended, but in a watchful expect-ancy, a balanced poise of contemplation. We hibernate, lazing and pottering through the brief hours of daylight, and then retir-ing to enjoy those long studious evenings of the north, which gave the ancient Icelanders, battened snugly down for six months, the chance to hang up their tools and weapons, and write poems about the things they did in summer.

The days were very short; both milkings were done by lamp-light, and the cows had a bare six hours at pasture. As our tenancy was so soon to come to an end, we did not trouble to do any work that was not strictly necessary. We went about the home-stead, sawing logs, chopping kindlings, sorting potatoes, making our food and eating it; we carried hay to the byre and sheaves to the Belties in the wood. There was no hurry or fuss. We had time to gossip with Mr Gordon at the wall of the poultry field, or with the shepherd in the cart-shed, and leisure to be more consciously aware of the beauty of our surroundings. Not that we ever forgot it for long, even in our busiest hours; it lay at the heart of life, and if it seemed not always fully perceived or openly appreciated, it was because of its very depth and nearness, like daily bread, which we do not notice much until it is taken from us.

In these days the cows were grazing in the sawmill field, which had no water, and at nightfall we would drive them down to the burn for a drink. Just below the turn of the farm road the water spread into a broad shallow pool fringed with sedges and lush grass and the flat green blades of the seilisdeir or wild iris. Here the cattle would stand and drink, some deeply, others in careless mouthfuls, butting their neighbours or picking at the grasses on the edge; one or two would stray on to the boggy slope beyond, and start to graze there. The two dogs sat alert at the roadside, waiting with many a loud impatient yawn for the signal of departure. We stood leaning on our sticks, watching the cows with bodily senses while our thoughts strayed like sheep turned loose upon the hill. There was no wind: night was near, and the world with its living freight seemed to drift into silence, as a ship becalmed moves with the tide. Snow lay on the highest Cuillins, lending them a strange and inaccessible majesty. This chain is nowhere much above 3,000 feet, but rising straight from sea-level without intervening foot-hills, it gives in certain conditions of light and atmosphere an illusion of height and remoteness unequalled by far loftier mountain ranges. So it was that evening. Between base and summit extended a horizontal veil of grey cloud, above which the snowy rampart stood clear, gathering to itself whatever light still lingered in the sky. Thus disconnected from its roots, in a world where measured distance was abolished, it looked immensely high and far away, like those undiscovered peaks men dream about and throw away their lives to win. Below the mountains lay a darker world, in which night had already fallen. The open fields and hillsides, dun with crumbling fern and withered grass, wore a subdued and pensive air, as if resigned to some inevitable conquest. On the far side of the burn was a solid stone wall, guarding a tract of massed and premature gloom. Tall trunks of pine and fir rose straight like pillars, their tapering tops lost in an invisible network of branches. Some of the outer trees were dead, and their barren stems gleamed naked in the twilight, but inside the wood was darkness profound and triumphant. Against the barrier of this silence the burn, shrunken by weeks of scanty rainfall, raised its frail, insistent tinkle. Down in the oak-woods an owl hooted; far away in the valley a crofter's collie was barking home the cows.

We stood there leaning on our sticks, as herdsmen stood in the days of Abraham, watering their beasts at the precious streams of a thirsty land, and thinking the long slow thoughts of all watchers of flocks since the beginning of the world. A light sprang up in the Gordons' cottage, and the plume of smoke thickened as fresh logs were piled upon the fire. Gathering our herd, we turned them up the brae. They walked slowly in single file; the dogs with lolling tongues and tails waving ran in and out among that forest of indifferent legs, barking here and nipping there, and often gazing up in admiration at the vast and massive bodies of their charges, so maddening in their slowness but impressive with sheer bulk and weight. Then through the gate and into the byre, where a rattle of chains and the rending sound of straw torn greedily from the racks proclaimed the finish of another day.

On the West Coast, long periods of severe weather are very rare. There will be a day or two of sharp frost, bringing a thin layer of ice to the lochs, then a rapid thaw, with so swift and sudden a rise of temperature that everyone feels slack and ener-vated as though by great heat. Rain will turn to sleet, then to snow, which will lie for a few hours, or even a day or two; then the wind will shift to the south, the snow vanishes as if by magic, and the world turns to slush. Another shift of wind, with a clear calm night: back comes the frost, and the roads become glass. And so forth in a perpetual see-saw till the coming of spring.

One Sunday in January the snow lay three or four inches deep in the steading. The sun was bright, but there was a woolly look on the horizon that seemed to promise more. By midday the Belties, thin and empty-looking, were bunched at the gate of the wood, bawling for sheaves. The byre cattle were wandering among the rushes, picking at the sharp reddish-brown spikes that rose above the level of the snow. On the ruffled iron-grey surface of the loch, a pair of wild swans floated; their gleaming plumage looked very dirty in comparison with the dazzling whiteness of their surroundings. In the stable field, Dick was scraping with his forefoot to clear a patch for grazing. As I walked up the brae, bowed under a load of sheaves, with eyes on the ground, I was fascinated by the tracks in the snow, which made the road like an open book. Tracks of hens, of hoodie crows, of dogs, cattle, and rabbits, the hob-nailed soles of the shepherd, Peter's worn rubber boots, the tyres of the post's cycle. The Belties saw me coming, and let out a salvo of bellows. I loosened the bands of the sheaves and threw them over the gate. Strewn on the snowy ground, or projecting from the black muzzles of the heifers, they shone with ruddy gold. I thought of the hot weary day on which they had been cut and bound, of the weeks they had stood soaking in stooks and hand-ricks; and here they were rustling with dryness, vanishing like smoke before the eager onslaught of hungry cattle. I turned to go. The Sound was crisping darkly under a fresh breeze; its shores for about fifty feet above sea level, as well as the smaller islands, were clear of snow and looked as black as ink beneath the dazzling slopes above. Large billowing clouds, bright with the peculiar hard brilliance of cold, sailed overhead; the farther distances were vague and woolly, with here and there a faint smudge of falling flakes. Clumps of tawny weathered rushes, casting blue shadows on the snow between them, glowed with a strange intensity of colour. Since we had seen so many curious types of weather, I wondered if this were the prelude of an unusually long spell of winter; but the same night there was a change of wind, the snow became soft and pitted, and vanished next day in a sea of mud. On Tuesday the Belties were lying in the green grass, chewing their cud and sniffing the soft balmy air. The place was running with water, and fearing the effect of a sudden frost on the road, I sent the cart to the station for stores. I need not have bothered. The air remained moist and mild as if it would never freeze again.

But you cannot be sure. In 1929 we had a hard frost which lasted for the greater part of January. The deepest lochs, even those fed by springs, were frozen hard, and in estuaries and sheltered bays the sea itself was rimmed with ice. The fan-shaped oozes on the hill looked like small glaciers, and water dripping from the face of the crags formed icicles, some of them twenty or thirty feet long, which when the thaw came lost their hold and crashed down upon the road in shattered blocks. Before the frost set in there had been much rain, so that the burns were full and everything saturated with moisture. But frost means drought; the moisture evaporated, leaving many a rush or plant-stem englobed in transparent ice, like some rare object in a glass bottle. The burns were frozen over at their original level; but as the water began to fall, a space was left beneath the canopy of ice, in which the tinkle of running water was magnified and given a definite musical value. Had these strange sounds been heard in some more populous place and at a milder time of year, you would have imagined that some musical instrument was being practised – perhaps a ukelele. Another curious musical illusion was caused by the sinking of loch water under the crust of ice, which unsupported would crack at intervals, and emit a curious bell-like note, which (as far as I could test it by memory with the piano at home) was the D-natural above middle C. The dogs heard it also and showed much uneasiness, and I do not wonder; for in the midst of a stillness as profound as any known in Nature there was something alarming about the intrusion of this rare sound, which I have never heard before or since. The familiar hill was changed to something new and rather disquieting. There was no cry of bird or hum of insect to break that charmed silence. The friendly sounds of daily life had ceased, leaving us nothing to hear but the void behind them; and we wondered if it was the voice of silence itself that we heard, or some physical buzzing in our own ears.

Our life was dominated by bags. We carted them from the station, and hauled them up the granary stairs – bags of meal for the hens and bruised oats and cake for the cattle, heavy unwieldy things with that exhausting dead weight that makes a lightish bag harder to carry than a full-grown man. The granary stairs were steep and rotten, and at a vital point in the descent there was a tread missing; and when we went up on a dark night or in a gale of wind, on wooden steps slimy with moisture or with snow half frozen or half melting, we had to watch our feet and cling to a far from reliable handrail. There is also an art of handling bags, and of filling or emptying them, which altogether eludes me. To transfer the contents of one bag into another without spilling, cursing, or wasting time seems to me a special gift, like absolute pitch or speaking with tongues; and even if it is not quite that, I am certain that I shall never acquire it, not if I spend the rest of my life in a granary. Most of these bags were no longer capable of holding anything but other bags, as they had been holed by rats and mice, on which neither dogs, traps, nor poison seemed to have any effect. Herself would have said that this was a judgement on us for our judicial murder of Charlie; but I am afraid that even in his day there were a good many rats and mice in the granary.

I have rather a weakness for mice. They spoil good things, like food and sleep, and reason tells me that they should be exterminated; indeed there is a tin of Rodine in my kitchen cupboard. But when they dart out of a hole like a flash of darkness, and sit with their soft fur and twiddling whiskers, looking at me with bright beady eyes, I cannot resist them. There was one fearless little fellow who would run all over the kitchen in daylight, when it was full of people and dogs; he would even play about my feet while I wrote. But one night Thos saw him, and by some fluke, for he was a clumsy hunter, managed to catch the mouse; it was seized, killed, and bolted, skin, bones and all, in a moment of time. I never scream at mice, and at one time I used to pride myself on my fortitude, thinking that it would prove me a fit mate for some pioneer of empire. But the Laird soon disillusioned me. Far from being an attractive or endearing quality, he said, it was just the opposite; in fact, if I had screamed at mice, I should no doubt have found a husband. But even so, I cannot bring myself to scream at them – they are too soft and pretty. There were often nests of field mice at the bottom of the long grass in the hay-field, which as the swathes fell were exposed to view. Murdo and Willie, with the delight in promiscuous slaughter that most male creatures seem to have by nature, would trample on the young ones and throw them to the dogs. I hated them for it, but perhaps they could not help it. Nearly all small boys, if left to themselves, kill everything within range at sight, and big boys are not very different from small ones.

Soon after the New Year the whole neighbourhood was convulsed by the news of a burglary at Rattray's. He and his wife had gone away for a month's holiday, having first sold their dog and transported their score or so of hens, houses and all, to be fed and tended by the shepherd, who had also been instructed to keep an eye on the house. Going up one morning he found a small window at the back broken and three tins missing from the window-sill – one of pineapples, one of tomato soup, and one of pears! He came down to the farm in great agitation, announcing that he would go straight to the village and send a phone to the 'pollis'. The sergeant, who was new to the district, came promptly on a bicycle, inspected the scene of the crime, questioned everyone at great length, and took particulars of the missing tins. Next day Mr Gordon wanted to know if we had eaten any tinned pears lately. As it happened, we had; and producing a tin, which he had found on the dung-heap, he asked me to identify it. Now the tin he showed me was not of the brand I favour, and had not been opened with a tin-opener, but the lid was gashed across as if by a blow from a hatchet. Also, the official tin-dump at Achnabo was elsewhere. The thief must have come down the road with his booty, opened the tin of pears with our hatchet, which is kept in the cart shed, drained the contents and flung the tin on the dung-heap opposite, perhaps forgetting that we do not throw our rubbish there, or perhaps (dark thought) wishing to cast suspicion on Peter! On his next visit the 'pollis' was shown the tin and took it away, perhaps to send to Scotland Yard. He would have liked the hatchet too, but we could not spare it. He came several times, asking questions and taking notes, and at last apologised for his importunity. 'Ach, well, it will give you something to do', I said. At which he bridled, and replied coldly that he had plenty to do without that. We had a shrewd idea of the thief's identity, and so had everyone else; but there was no evidence, and we left the Pollis to pursue his investigations without giving any further clues. The thing was plainly done as much to annoy as for the sake of plunder, for the total value of the stolen goods was under four shillings! The Laird heard that the house had been broken into and completely ransacked: so much for the power of rumour.

Every Sunday evening we went to supper at Strathascaig. For a whole year and two months we had not missed a single week; it was our only diversion, and we prized it accordingly. The first winter we mostly took the trap; but the time and fatigue saved in driving was counterbalanced by the trouble of yoking and unyok-ing, the badness of the road, the dimness of our lights, and the uncertainty of Dick's behaviour. We had to catch the horse, feed him, harness him, pull out the trap, and yoke him into it. Then drive in twilight or complete darkness over four miles of unspeak-able road, ending in a rotten bridge without rails, a rutted track across a bog, and a ford of uncertain depth; and then unharness the horse, feed him, and tie him up in a strange stable beside an unfriendly mare. The trap was fitted with carriage candles which, though giving a little more light than the bedroom variety, were effectively dimmed by the mud spattered on the glass. They illu-minated a small space just in front of each wheel but shed no light on the road beyond. Also one never knew how long they would last, and the adjustment of a fresh candle in a high wind was no easy task. On wet or cold nights, any prolonged sitting in a high, exposed dog-cart was miserably chilling; our harness was much mended with string, and one of the shafts, which had a bolt worked loose, was lashed with rope to prevent excessive play. Since the cart accident, there was no telling what Dick might find to shy at; and we were not sorry when the bottom of one of the lamps was lost on the road or in the ford, and we could no longer drive at night.

The road was private, and belonged to two proprietors, each one of whom was waiting for the other to begin repairs. While they waited, the forces of Nature were combining to bring the road to a state of primeval chaos, in which the word 'repair' would cease to have any meaning. From Achnabo to Strathascaig the road passed first under trees in curious corrugations, pitted with deep potholes that in the driest weather were always full of water, and thence in a level stretch beside a freshwater loch, where the potholes were even deeper and more numerous. Some of these had been filled up with a kind of muddy clay, at the expense of a lorry and two men, but the clay was far worse than the water. This section of the road was open and windswept; we once encountered a gale there which blinded us with horizontal rain, nearly tore the reins from our grasp, and threatened every moment to upset the trap. From the loch the road descended a long hill, its surface nothing but a mass of small stones and rubble like the dry bed of a stream, for the loch had once overflowed, and the debris left by the flood had never been removed. From the foot of this hill there was a mile of quite good surface until about 300 yards above Murdo's township, where the road once more became a watercourse with stones and deep chasms, a condition caused by choked ditches and nothing else. After that the potholes began again, with overhanging and ingrowing trees, ending with a level flat where water covered the road completely, and beyond this a stony confusion made by a landslide which had been imperfectly cleared away. After that were more pot-holes, chasms, and over-hanging trees, and at intervals large stones that had crashed down from the crags. Then came a stretch of good surface, but the wind-ing road was so narrowed by encroaching whin bushes that the trap could not pass without getting scratched and it was impossi-ble to see anyone coming. And last of all we had to face the rotten unrailed bridge of turf and sleepers that spanned a tributary of the Strathascaig river, and the rough boggy cart-track and the ford with its hidden stones and iron gate quite often closed against us.

For three-quarters of its length the road passed under trees, most of which had shallow roots, since the thin and rocky nature of the soil prevented them from striking deeply down. In bad weather there was always a danger of fallen trees blocking the way, and I confess that I never cared to drive there when a strong gale roared through a thousand rocking branches. One evening in January, as the early dusk was beginning to fall, we turned a corner and found our way barred by a solid cliff of timber about fifteen feet high and I know not how broad, for five or six huge trees, mostly spruce and Lawson cypress, seventy to eighty feet high, had come down in a row like a pack of cards, torn from their shallow hold by a single gust of a rare northerly gale. Fortunately we were on foot and were able to climb round the top of the obstruction, which remained for many days unmoved, because no one wanted to be the first to tackle so big a job. Often I hear in fancy, loud above the racket of the storm, the prodigious rending crash of their collapse; with their spreading branches they must have covered a hundred yards of road. Lucky for us that we were not passing then! At times of heavy rain, the whole road was one great watercourse, and walking more like wading; the faint beam of the hurricane-lamp showed spreading pools or hurrying streams of muddy water, brimming in chasms or eddying round stones, the whole track being no more like a road as the towns-man conceives it than Dick's turnout resembles a smart saloon car. But I would not exchange either of them.

In these weekly walks we saw every kind of weather, and the

kinds were many; a whole book could be written about them alone. Since there were so many trees to block the vision, the nights of calm and stars were less memorable than those of wind and rain. On Christmas evening we were rounding the last corner, where the road emerges from the woods into the open glen of Strathascaig. The mild air was thick with drizzle, and full of the noise of a south wind which under the lee of the hill was only felt in fitful gusts. Our lantern, fringed with shimmering raindrops, had little strength to pierce the darkness. Below, the lights of Strathascaig, infinitely far away, blinked through swaying trees. Presently we were deafened by the rush of water from a burn in flood, a tributary of the main river, which poured down its rocky gorge and then stormed out into the open, churning the boulders in its stony bed, and foaming through the narrow archway of the bridge. Leaning over the parapet we swung the lantern forward and saw a dark swirl of water, flecked with foam and bubbles, and long grasses on the bank, limp and flattened like a drowned woman's hair, showing the recent level of a spate. There was a seedling whin in flower, jerking spasmodically as the water tore at its roots; below it was a gleam of metal and coloured paper from a tin thrown out by the shepherd.

We took the cart-track to the farm, which ran down a steep slippery brae to an unrailed bridge of railway sleepers covered with earth. At this point the burn changed its note; it lost depth and sonorousness, it fussed, jabbered, and snarled. The channel was wider and shallower, guarded on the down side by a low embankment which in many places was now breached and crumbling. In the old days, when money was plentiful and labour cheap and abundant, someone had the notion of diverting the burn, so that its former course might be occupied by a direct road to the steading across the ford. But as the new channel ran at a higher level than the old, there was a continual seepage of water, which made the intervening pasture a perpetual bog and the road itself an occasional river. We followed a rough footpath beside the burn. Here, under the lee of the southern slope, the air was perfectly still; and wherever the path bent away from the hurrying water we discovered that its persistent clamour was only one among a hundred other noises of the night. The rush and babble of a dozen small streams in spate, the groaning of the oak wood behind the steading, the hoarse roar of the big falls on the river, the deep and awful baying of the wind as it strained against the northern barrier of the glen: all these sounds, combined yet separate, rose and fell like the movement of a vast orchestra. We stood listening. The whole world was drowned in gloom and vapour; sight, touch, and smell were non-existent. The body was all ear, in strained attention. The darkness split before us; a fiery jet of steam, trailing a luminous wake, passed across the field of vision, and was lost. No comet, portending the crash of empires, could look more grim. The evening train was crossing the bridge; we must hurry or we should miss our dinner. We looked back. Behind us was the shoulder of a hill, up which the station road climbed in a slow ascent, and then dipped sharply out of sight. Suddenly from beyond the ridge a beam of glory shot fanwise into the air, becoming every moment larger and brighter. Then on the crest appeared two blazing orbs, like the eyes of some fabulous tiger. The fan-shaped vision faded, and instead long shafts of light played upon the hillside opposite. A car – probably the doctor's.

We crossed the long wooden footbridge, with its slippery planks and rotten hand-rails. Someone inside the house switched on the powerful lamp that was fixed to the gable for the convenience of outdoor workers. Strange paradox, that this farm, where hay and corn were cut with the scythe, should be equipped with electricity! The light was fierce and harsh, casting black and unfathomable shadows. There was something sinister about it, as if a flash of lightning had been arrested and imprisoned in a bulb. Opposite the gable was a wild cherry tree, twisted with age, spreading itself in a complicated network of branches and twigs. When the lamp was switched on, each minutest detail of this twiggery was revealed, delicately etched upon an inky background. Misty rain whirled past the globe, like golden dust thrown across a gulf of blackness. The courtyard of the byre was paved with cobblestones islanded in a sea of liquid mud. In the middle stood the cart, a battered, venerable relic, resting upon its shafts. Round it were grouped the roosting ducks and geese; at our approach their heads were untucked, and they gobbled softly and suspiciously. The firm contours of the wheels, the birds' spotless plumage, the ancient stone wall of the byre, the wet slates, the dark abyss of sky behind, were picked out in hard, uncompromising black and white, like a modern wood-engraving.

We went in by the back door. Often we were wet to the skin, and stood dripping on the threshold, with pools of water round our feet. But tonight we were only damp, and shedding our waterproofs and shoes, retired to change our clothes and appear as radiant guests at the Christmas feast. As I sat at the polished table, with its sparkling glass and silver, dressed in blue velvet and consuming with infinite gusto the choice food and various wines, I thought that only a fool would choose to fare sumptuously every day, since the pleasures of the table, so vivid and glorious when they come not often, are dulled by repetition till we eat and drink mechanically, and might just as well be served with the crofter's oatmeal and salt herring. For a whole year we had looked forward to this banquet, especially during the last few weeks, when we had been without butcher's meat; and now it had come, and we walked in paradise – rather a low Valhalla, it is true, but paradise enough for us. Full of good fare and warmed with wine, we wandered into the gaily decorated sitting-room and sat by the blazing fire, enjoying good fellowship and talk and the genial hospitality of our friends. And later, when cups of strong coffee had been served, we managed to play a fairly intelligent game of bridge.

At half past ten Thos thought it was time to go home. He rose and stretched elaborately, opening and closing all his toes in order; then with a loud and cavernous yawn he walked to the door, inviting us to follow. We got on our feet and began to take our lingering farewells. He wagged his tail, barked once or twice, sneezed three times, and yawned again. The door opened, and the dog rushed eagerly into the night, followed by his reluctant owners. We were replete and very sleepy, and the four-mile tramp home seemed rather long. We took a short cut over a part of no-man's-land that had once been ploughed and no crop taken; the old hard-packed furrows were bad to walk over in the dark, and a straight course difficult to steer. The wind groaned in the branches, but we were in shelter and felt little. At the loch-side we came out into the open, and a tearing gust struck us. It was too dark to see the water, but in the lulls we heard the short waves viciously slapping the stones that fringed the shore, and the hoarse roar of the gale in the trees on the farther shore. Our lamp dipped and flickered, but it kept alight until we reached the stead-ing. It was midnight, and the Gordons had long since gone to rest, leaving hot bottles in the beds of the strayed revellers. At this vacant hour, the cart-shed, with the trap resting on its shafts, the straddled sawing-horse and piled blocks and implements, had a strange, frozen look of action suddenly arrested like a busy town overwhelmed by some unforeseen disaster. In the byre, some wakeful cow rattled her chain, bringing the whole place back to life again. We walked over to the house. The door was unlocked, for we had lost the key long ago. The kitchen fire was out, and the place wore the quiet expectant air of an empty house that waits all night for its owner's return. We groped for matches and lighted a candle; the thin flame drew up, guttering in the draught, and revealed a batch of newly baked loaves and a jug of milk. A mouse, surprised on the hearth, gazed at us for a moment with bright eyes and vanished down a hole. It was a squatter, tolerated only because we were too lazy to set traps and poison every night. Somewhere behind the wainscoting it had a permanent home, and a nest lined with tiny fragments of *The Times*.

As for us – where should we be next Christmas night, and who would be here in our place? The wind, moaning round the gable, gave no answer; in the dark world of Nature outside the door, that question had no meaning. The gravest humans debating their fate were of no more account than a mouse twiddling its whiskers on the hearth. A squall rushed out of the south, and set the trees swaying and rushes bending, as they had done since the beginning of the world. Through a rift in the scudding clouds a few stars looked down on the homestead with remote, indifferent glance, as they would look a thousand years hence, when all these stones were crumbled into dust. What would it matter by whose hands the harvests were gathered, or even if they were gathered at all? The forces that raised and ripened them, the sun and rain in their circling seasons, would still be busy, and in the fullness of time there would perhaps be someone else to work at Achnabo and find pleasure in his labour.